THE DESCRIPTIVE AND PHYSIOLOGICAL

ANATOMY

OF THE

BRAIN, SPINAL CORD,

AND

GANGLIONS,

AND OF THEIR COVERINGS.

ADAPTED FOR THE USE OF STUDENTS.

BY

ROBERT BENTLEY TODD, M.D., F.R.S.

FELLOW OF THE COLLEGE OF PHYSICIANS, PHYSICIAN TO KING'S COLLEGE
HOSPITAL, AND PROFESSOR OF PHYSIOLOGY IN KING'S COLLEGE, LONDON.

LONDON:
SHERWOOD, GILBERT, AND PIPER,
23, PATERNOSTER-ROW.

1845.

TABLE OF CONTENTS.

	PAGE
APHORISMS RESPECTING THE NERVOUS SYSTEM	ix

CHAPTER I.
Description of a nervous centre	1
Coverings of ganglions	3
Meninges	3
Dura mater	4
The spinal dura mater	4
The cranial dura mater	7
Processes of the cranial dura mater	8
Arteries and veins of the dura mater	10
Spinal sinuses	11
Cranial sinuses	14
Spinal pia mater	23
Cerebral pia mater	25
Choroid plexuses of the lateral ventricles	26
Velum interpositum	28
Choroid plexus of the fourth ventricle	28
The arachnoid	32
Spinal arachnoid	33
Cerebral arachnoid	35
The cerebro-spinal fluid	38
Variations in its quantity	50
Analysis of it	53
Glandulæ Pacchioni	55
Are they natural or morbid structures?	58
Ligamentum dentatum	60

CHAPTER II.
General remarks on the structure of the nervous centres	63
Of the vesicular or grey nervous matter	64
Caudate nerve-vesicles	66
Development of the vesicular matter	69
Pigment in the vesicular matter	72
Of the structure of ganglions	73

CHAPTER III.
Of the cerebro-spinal centre in general	77
Of the spinal cord	77
Its fissures	82, 83
Grey commissure	83

CONTENTS.

	PAGE
Subdivision of the cord	85
Its columns	86
Its grey matter	87
Central canal	93
The bloodvessels of the cord	97
Of the spinal nerves	99
Their origin	105
Mr. Grainger's observations	106

CHAPTER IV.

The encephalon	111
Its weight	113
Reid's observations	114
Tiedemann's observations	116
Of the brain in different races of mankind	122
Subdivision of the encephalon	128
Modes of dissecting the brain	129
Willis's method	131
Surface of the encephalon	137
Base of the brain	139
The fissure of Sylvius	139
Dissection of the brain from above downwards; its topography	145
The lateral ventricles	147
The fornix	151
The third ventricle	153
The pineal gland	155
The mesocephale	157
The cerebellum	159
The fourth ventricle	160

CHAPTER V.

Of the medulla oblongata	161
Signification of the term	161, 162
The anterior pyramids	164
Decussating fibres	165
Arciform fibres	166
Crossed influence of the brain	169
Restiform columns	171
Posterior pyramidal columns	174
Olivary bodies	175
Corpus dentatum	175
Olivary columns	177
Interpretation of the columns of the medulla oblongata	178
Nerves of the medulla oblongata	179

CHAPTER VI.

Of the mesocephale	180
Pons Varolii, or annular protuberance	181
Tubercula quadrigemina	183
Iter	185
Analysis of the mesocephale	186–7

CHAPTER VII.

Of the cerebellum	188
Its subdivision	189
Its weight	189
The valley	192
The hemispheres	193
Lobes	194
Superior vermiform process	197
Inferior vermiform process	198
Posterior medullary velum	191
Structure of the hemispheres	202–3
Corpus dentatum	204
Of the fourth ventricle	205

CHAPTER VIII.

Of the hemispheres of the brain	208
Of the convolutions	208
Convolutions in the inferior animals	212
Primary and secondary convolutions	215
Insula of Reil	216
Want of symmetry in the convolutions of the hemispheres of the human brain	217
Constant convolutions	218
Hippocampi	221
Corpora striata and optic thalami	223
Structure of corpus striatum	225
Corpora geniculata	228
Structure of optic thalamus	229
Corpora mammillaria	230
The commissures of the brain	231
Tuber cinereum	237
The pituitary body	237
Of the ventricles of the brain	239
Membrane of the ventricles	240

CHAPTER IX.

Of the circulation in the brain	241
Arteries of the brain	242

	PAGE
Circle of Willis	243
Veins of the brain	245
Venæ Galeni	245
Can the absolute quantity of blood in the brain vary?	247
Delirium in anæmia	249

CHAPTER X.

Origins of the encephalic nerves	250
Willis's arrangement	251
Olfactory lobes and nerves	251
Optic nerves	255
Optic tract	255
Optic chiasma	258
The third nerve	260
The fourth nerve	260
The fifth nerve	261
The sixth nerve	263
The seventh nerve	263
Portio dura	263
Portio mollis	264
The eighth nerve	264
The glosso-pharyngeal nerve	265
The par vagum	265
The spinal accessory	265
Concluding remarks	267

CHAPTER XI.

Microscopical anatomy of the spinal cord and brain	269
Grey matter of the cord	270
Stilling's researches	271
Structure of the medulla oblongata	273
Structure of the convolutions	276
Layer of light matter in the grey substance of certain convolutions	276

CHAPTER XII.

Hypothesis of the action of the brain	278
Mechanism of a voluntary act	279
Mechanism of sensations	280
Influence of emotions	282
Action of the cerebellum	283

PREFACE.

The following pages are a reprint of the article "Nervous Centres" in the Cyclopædia of Anatomy and Physiology, with several additions, and some alterations suggested by its re-perusal as it again passed through the press.

The Author has been induced to submit this book to the public in its present shape, by the frequent complaints which have reached him from students and others, of the want of a description of the brain and spinal cord, embodying the most recent observations on the anatomy of these important organs.

He is not vain enough to suppose that he has succeeded in completely supplying this defect. On the contrary he is conscious of many imperfections; he has, indeed, found it necessary to leave many points incomplete, in consequence of the great difficulty of coming to a decisive conclusion respecting them, preferring to incur the charge of having given a superficial description, to advancing a positive statement on insufficient evidence. In the composition of the work he has relied chiefly on his own dissections, and the descriptions are the same as he has been in the habit of giving for many years in his lectures.

The Author has been anxious to render this anatomy of the brain as physiological and as practical as the nature of the subject would admit. But he has purposely abstained from physiological discussions, as foreign to his object, and tending to divert the attention of the reader from the anatomical inquiry.

It will be remarked that he has studied conciseness of description and has avoided much reference to the opinions of others; being strongly impressed with the conviction that the frequent reference to merely hypothetical views has greatly impaired the clearness and precision of some, otherwise excellent, anatomical and physiological writings. The Author, therefore, contents himself with expressing here his obligation to the following works: Chaussier sur l'Encephale; Cruveilhier Anatomie Descriptive; Hildebrandt Anatomie; Reil's Essays translated by Mayo; Mr. Solly's Work on the Brain; Foville Anatomie du Système Nerveux; Leuret Anat. Comp. du Syst. Nerv.; Longet Anat. et Physiol. du Système Nerveux.

26, *Parliament Street,*
September, 1845.

APHORISMS

RESPECTING THE NERVOUS SYSTEM.

1.—The nervous system consists of internuncial chords or nerves, and of centres, namely, of ganglions, the spinal cord, and the brain.

2.—There are two kinds of nervous structure, the white or fibrous, and the grey or vesicular.

3.—The fibrous matter consists of fibres of two kinds, the *tubular* and the *gelatinous*. The vesicular matter is composed of vesicles, surrounded by a granular matter which ensheathes and connects them; these are of two kinds, the *simple* and the *caudate*, (*vide* p. 65.)

4.—In the centres the vesicular and fibrous structures are united: in the nerves the fibrous matter alone exists.

5.—The union of the two kinds of nervous structure in the centres is necessary for the generation of the nervous force, and for the maintenance of the nutrition of all parts of the nervous system.

6.—Nerves are of two kinds, motor and sensitive. In the motor nerves the nervous force ordinarily travels from the centre to the periphery, in the sensitive nerves it travels from the periphery to the centre. The former is therefore *efferent*, with reference to the centre, in its ordinary mode of conducting the nervous force; the latter is *afferent*. This difference is to be attributed, not to any intrinsic peculiarity of structure, but to a difference in the organization of each class of nerves at the periphery and at the centre.

7.—Motor nerves are connected at their central extremities with the centre of voluntary motion, and at their peripheral extremities with muscles. Sensitive nerves are connected at their central extremities with the centre of sensation, and at their peripheral extremities with sentient surfaces, or with particular organs adapted to the reception of certain agents,—as light, sound.

8.—The stimuli by means of which the nervous force may be excited are of two kinds, **mental** and **physical**.

9.—The ordinary stimulus of motor nerves is mental—the will; but a physical change in a nervous centre, or in the nerve fibres themselves, may excite these nerves.

10.—The ordinary stimulus of sensitive nerves is a physical impression, the change excited by which is perceived by the mind. A physical change in the course of the fibres of the nerve, or at their central extremity, may give rise to a sensation similar to that ordinarily produced by an impression at the periphery. Or the mind may excite a portion of the centre of sensation, so as to occasion a feeling, of precisely the same nature as that, which usually results from an external cause.

11.—The fibres of nerves remain distinct, but in juxta-position, from centre to periphery; no anastomosis takes place between them. An interchange of fibres does occur, however, between neighbouring trunks of nerves, by means of which different portions of the nervous centres are brought into connection with each other, and with an extended peripheral surface.

12.—The nervous force is a polar force, resembling electricity in the instantaneousness of its development and in the rapidity of its propagation, but differing from it in several important features.

13.—It is a peculiar feature of the nervous force, not only that it may be developed under the influence either of a mental or of a physical stimulus; but that, being excited by a physical stimulus, it is capable of affecting the mind.

14.—Hence nervous actions may be conveniently distinguished into two classes, according as the mind participates in them or not; namely, mental and physical.

15.—Mental nervous actions are those which originate in and are excited by an act of the mind, as all voluntary actions; or which, originating in a physical impression, produce an affection of the mind, as in all sensations.

16.—Physical nervous actions are those which take place without the *necessary* intervention of the mind, and which result from a physical exciting cause. Of this kind are all those actions which are caused by a physical change, frequently morbid, in a centre or in a nerve, and those which result from the excitation of the vesicular matter, in which a motor nerve is implanted, by the stimulation of the peripheral fibres of a contiguous sensitive nerve. These latter have been designated *reflex actions*, and, although most of them, in health, are attended with consciousness, still this mental state is in no way necessary to the perfection of the nervous act, and should be regarded rather as an incidental accompaniment to it than as an essential part of it.

THE BRAIN, SPINAL CORD,

AND

GANGLIONS.

CHAPTER I.

Description of a nervous centre—Ganglions, the spinal cord, the brain—Coverings of ganglions—Meninges—Dura mater—Bloodvessels of the dura mater—Pia mater—Arachnoid membrane—Cerebro-spinal fluid—Pacchionian bodies—Ligamentum dentatum.

A NERVOUS CENTRE may be defined as a mass composed of grey and white nervous matter with which nerves are intimately connected. In a physiological point of view it is à centre of nervous action, as nerves appear to conduct to it as well as from it.

The nervous centres in the human subject are the GANGLIONS, the SPINAL CORD, and the BRAIN.

The ganglions are small masses occupying certain situations in the body. They are extremely numerous in the human body, and very variable in shape and size. One great subdivision of them, in man and the mammalia, is connected with the posterior roots of the spinal, and with certain encephalic nerves. Another class belongs to the sympathetic system. In the Invertebrata the nervous system is made up of a series of them variously disposed, with their afferent, efferent, and connecting nerves.

The spinal cord and the brain are peculiar to the great class of vertebrated animals. They may be regarded as compound ganglions, being physiologically resolvable into a series of smaller centres, which are, to a certain extent, independent of each other. Viewed anatomically, they are not so obviously divisible: in the spinal cord, in which the independent influence of separate segments may be most easily demonstrated by physiological experiment, no anatomical subdivision is obvious, for the segments are fused together into a cylindroid body, which has a certain relation to the length and muscular activity of the animal. Indications, however, of this composite form of the spinal cord are afforded, in the marked difference of dimensions which certain parts of it present when compared with others, there being always a manifest correspondence between the size of any segment of the cord and the motor or sensitive endowment of that portion of the body which receives its nerves from it. And the case of the common gurnard (*Trigla Lyra*) may be here quoted as a remarkable instance of the developement of distinct gangliform bodies on a portion of the cord, in accordance with a particular exaltation of tactile sensibility.

The brain is much more evidently made up of a series of separate centres or smaller masses, exhibiting sufficiently distinct boundaries on their surfaces, but so intimately connected by what are called *commissural* or uniting fibres, as to manifest the same kind of fusion (although to a less degree) as that noticed in the spinal cord. These gangliform bodies are so readily distinguishable from one another, that from the earliest periods of anatomical investigation each of them has been designated by a distinct name, which is generally derived from some prominent feature of the body itself, or from the name of some familiar object which it has been supposed (often fancifully) to resemble. The aggregate of these bodies is known in popular language by the name of *Brain*, (a word of Saxon origin, sometimes used in the plural); this word, however, anatomically speaking, is applicable only to the great hemispheric lobes which form the largest portion of the whole mass; and the term *Encephalon* may be more correctly used to denote the whole of the intra-cranial contents.

It is proposed in the following pages to consider the general and descriptive anatomy of these nervous centres severally, beginning with an examination of their coverings.

Coverings of the ganglions.—Every ganglion is covered by a more or less dense layer of white fibrous tissue, similar to that which forms the neurilemma of nerves. It performs the same office for the elements of the ganglions that the neurilemma does for those of the nerves; that is, it gives them a mechanical support, and is the medium through which bloodvessels are conveyed to their nervous matter. It is continuous with the neurilemma of the nerves which are connected with the ganglions. It is found in all forms and classes of ganglions, presenting the same essential characters. These bodies are generally surrounded by and imbedded in a considerable quantity of fat, which also involves more or less the nerves that proceed from them.

COVERINGS OF THE SPINAL CORD AND BRAIN.—These are also called the *membranes* of these centres, or the meninges (μηνιγξ, membrana). They are three in number. Those of the brain are continuous with those of the spinal cord, but, as there are certain distinctive characters proper to each, it will be convenient to describe the cerebral and spinal meninges separately. They are, enumerating them from without inwards, the *dura mater*, the *arachnoid membrane*, and the *pia mater*.

The term, mater, μητηρ, originated with the Arabian anatomists, who regarded these membranes as the parents of all others in the body. Galen adopted the word μηνιγξ, and distinguished the first and last of the membranes above enumerated by the adjectives παχυτερη and λεπτη. The Germans use the word *haut*, and designate these membranes as *hautige Hullen des Gehirns und des Ruckenmarkes; die harte Hirnhaut, die harte Ruckenmarkhaut*, the dura mater of the brain and spinal cord; *die Spinnwebenhaut*, the arachnoid; and *die weiche Haut*, the pia mater.

Dura mater.—The dura mater is a dense membrane composed almost exclusively of white fibrous tissue. It has all the characters, physical and vital, of that texture, possessing great strength and flexibility with but little elasticity. It is freely supplied by bloodvessels, and at certain situations, which will be more particularly described by-and-bye, it separates into two laminæ, which inclose prolongations of the lining membrane of the venous system, forming peculiar sanguiferous channels, which are commonly known by the name of *sinuses*. It has an apparent lamellar disposition, from the fact of its fibres being arranged in different planes. In the child a subdivision into two layers may sometimes be easily effected. Some nerves have been demonstrated in the dura mater; a branch of the fifth nerve has been particularly described and delineated by Arnold, as passing in a recurrent course between the laminæ of the tentorium, and Pappenheim has found nervous fibres in the cerebral dura mater derived from the superior maxillary division of the fifth, from the fourth nerve, from the vidian, and probably also from the frontal branch of the ophthalmic.*

The *spinal dura mater* is in shape adapted to the vertebral canal. It is a hollow cylinder, tapering somewhat at its lower

---

* Valentin Repertorium, vol. v. p. 87.

extremity to correspond with the sacral portion of the canal. It adheres very firmly all round the foramen magnum of the occipital bone. From thence it is continued down to the sacrum without forming any adhesion to bone. On the posterior and lateral surfaces it is covered by a layer of soft, oily, reddish fat, which intervenes between it and the inner surfaces of the vertebral laminæ and processes, and in these situations, as well as to a less degree in front, we find a very intricate plexus of veins, some of which are of considerable size. The fatty deposit is most abundant in the sacral region. In front the dura mater adheres by a close areolar tissue to the posterior common ligament, and here of course the adipose tissue is deficient. At the foramen magnum the continuity of the spinal dura mater with that of the cranium is distinct, and here, indeed, the former appears as a funnel-shaped prolongation of the latter; both are, in truth, portions of the same membrane adapted to the difference of shape of the nervous centres with which they are respectively connected.

On the sides the spinal dura mater is perforated by orifices which give exit to the roots of the nerves which arise from the spinal cord. When examined from within, these foramina are found to be arranged in pairs, each pair corresponding to the point of exit of a spinal nerve. The foramen which transmits the anterior root is separated from that which gives passage to the posterior one, by a narrow slip of fibrous membrane. These foramina are slit-like in form, taking the vertical direction. On the outer surface of the dura mater the distinction between them is not evident without dissection, for there the fibrous membrane being prolonged from the margins of the openings around the nerves, the sheaths thus formed coalesce and surround both roots. The number of these orifices is of course the same as that of the roots of the nerves which pass through the dura mater.

The internal surface of the spinal dura mater is perfectly smooth and moist in the healthy state, owing to its being lined by the parietal layer of the arachnoid membrane. In the intervals between the orifices for the transmission of each pair of spinal nerves, it receives the pointed attachments of the ligamentum denticulatum, to be described more fully by-and-bye.

It is evident from the preceding description that the spinal dura mater cannot perform the office of a periosteum to the osseous walls of the spinal canal, for at every point it is separated from them by texture of a different kind, and, moreover, the vertebræ are provided with a distinct periosteum. The prolongations of dura mater over the nerves at each of the intervertebral foramina serve to fix that membrane at the sides throughout the whole extent of the vertebral canal, so as to prevent its lateral displacement. At the lower extremity of the sacral canal the dura mater ends in a blunt point, and from this certain processes may be traced towards the coccyx. Of these the central one is continuous with the filiform prolongation from the pia mater, which is inserted into the inferior extremity of the dura mater, and is implanted into the last bone of the coccyx. The thread-like processes which are seen on each side are the sheaths of the last sacral nerves and of the coccygeal nerve, which pass some distance in the canal before they reach the foramina for their transmission outwards.

It is easy to convince oneself that the spinal dura mater is far larger than would be necessary for the reception of the cord. When the fluid immediately surrounding this organ has been suffered to escape, the dura mater appears quite loose, flaccid, and wrinkled. By blowing air or injecting water into its canal, it may be rendered tense again. This looseness of the dura mater is most conspicuous at its lowest part, in the lumbar and sacral regions, where it forms, as Cruveilhier says, " autour de la queue de cheval une vaste ampoule, qui parait n'avoir d'autre utilité que de servir de reservoir au liquide cephalo-rachidienne."

The dura mater adapts itself, in point of size, to the varying dimensions of the spinal canal in its different regions, which again appear to be influenced by variations in the dimensions of the spinal cord. Thus, it swells in the cervical and in the lumbar region, at both which places there are corresponding enlargements of the cord. Its most contracted portion is that which occupies the dorsal region.

Cranial dura mater.—The dura mater of the cranium differs in one leading circumstance from that of the spine,—namely, that it forms a periosteum to the inner surface of the cranial bones. We find it, therefore, very closely adherent to the whole interior of the cranium, and the free communication between the vessels of the dura mater and those of the bones serves materially to enhance the connexion between this membrane and the osseous surface. At some situations the adhesion is so very intimate that we experience great difficulty in attempting to separate the fibrous membrane from the subjacent bone. On the roofs of the orbits, the wings of the sphenoid bone, the petrous portions of the temporal bones, the margin of the occipital foramen, and opposite the sutures, the adhesion is very intimate.

This adhesion of the dura mater to the bones is found also to vary in degree at different periods of life. It is very intimate in old age, so much so that, in removing the calvaria, layers of bone often chip off, remaining in connexion with the fibrous membrane. In the adult, such a degree of adhesion as would give rise to this effect, ought to be regarded as morbid. In the young subject, while ossification and growth are going on, the adhesion is very intimate, so that in them great difficulty is experienced in removing the calvaria. Doubtless this intimate adhesion at this early period of life is due to the active share which the dura mater takes in conveying the material of nutrition and growth to the cranial parietes.

The cranial dura mater is not a simple bag. From its internal surface partition-like processes pass inwards, which serve to separate certain subdivisions of the encephalon. These are, the *falx cerebri*, the *tentorium cerebelli*, and the *falx cerebelli*.

The *falx cerebri* is a process of fibrous membrane corresponding to the mesial plane and lying in the great median fissure of the brain, where it separates the lateral hemispheres from each other. Its shape is falciform; its superior convex border corresponds to the frontal and sagittal sutures, and encloses the great longitudinal sinus; its inferior border is concave and much shorter than the superior, and corresponds to the superior surface of the corpus callosum. In front the falx is very narrow and almost pointed; it embraces the *crista galli* of the ethmoid bone, which appears to be enclosed between its layers. As the falx proceeds backwards it increases considerably in depth; its superior edge may be traced back to the internal occipital protuberance; its inferior edge terminates at a point corresponding to the middle line of the posterior margin of the corpus callosum. The falx cerebri contains within it, along its posterior border, a large vein, which is called *the inferior longitudinal sinus*.

The falx cerebri is continuous at its posterior border on each side with the *tentorium cerebelli*. This process is nearly horizontal in its direction; it forms a vaulted roof to a cavity (the floor of which corresponds to the occipital fossæ) in which the cerebellum is lodged. Its upper surface is convex on each side of the attachment of the posterior extremity of the falx cerebri: it supports the posterior lobes of the brain. The inferior surface is adapted to the upper convex surfaces of the cerebellar hemispheres. Its posterior and outer edge adheres to the occipital bone and to the posterior border of the petrous portion of the temporal bone, reaching as far inwards as the posterior clinoid processes of the sella Turcica. The occipital portion of this edge contains a considerable part of the lateral sinus (*fig*. 3, *e*), and that portion which adheres to the petrous bone contains the superior petrosal sinus. The anterior or inner margin of the tentorium is concave and free in the greater part of its extent; it forms the posterior and lateral boundary of a large opening (which the sella Turcica completes in front), through which the crura cerebri and other parts connected with them pass. This margin is attached by its anterior extremities to the anterior clinoid processes, to reach which it crosses the posterior border. The crossing of these two edges at a point external to the sella Turcica gives rise to the formation of a little triangular space, the base of which corresponds to the sella Turcica; its outer angle is perforated for the transmission of the third pair of nerves, and its anterior one for that of the fourth pair.

From the inferior surface of the tentorium cerebelli at its posterior edge, a short and thick fold of very slight depth descends to the posterior edge of the foramen magnum. This is the *falx cerebelli*; it corresponds to the median notch between the hemispheres of the cerebellum. Its anterior border is slightly concave. Two veins called *occipital sinuses* are contained in it.

The internal surface of the cranial dura mater presents the same smooth appearance as we have noticed in the spinal membrane of the same name. We observe, however, an exception to this on each side of the line along the great longitudinal sinus. Here it is very common to find a peculiar cribriform appearance, which occupies a space of from half an inch to two inches in length and not more than a quarter of an inch in breadth, but exhibiting great difference in various subjects as to the number and depth of the foramina or depressions upon which the sieve-like structure depends. These depressions are caused by the presence of little bodies, *glandulæ Pacchioni*, which grow from the membrane that covers the brain, and will be described by-and-bye. The anatomist may expect to find in a large proportion of adult brains a greater or less degree of adhesion between these parts of the dura mater and the edges of the hemispheres of the brain.

The dura mater is perforated by numerous orifices for the transmission of the encephalic nerves. It adheres firmly to the border of each of the foramina in the cranial bones, and is partly continued in the shape of neurilemma over the nerve that escapes through it. In the case of the optic nerve a strong fibrous sheath is prolonged from the dura mater, and at the same time that membrane appears to become continuous with the periosteum of the orbit, as if it had, opposite the optic foramen, split into two layers, one of which formed the sheath of the optic nerve, and the other applied itself to the interior of the orbit, forming a periosteum to the walls of that cavity.

Of the arteries and veins of the dura mater.—The disposition of the bloodvessels of the dura mater, both of the spine and of the cranium, deserves a special description. The former membrane derives its arteries from the numerous vessels which take their rise close to the spinal column in its various regions. These are ramifications of the abdominal and thoracic aorta or of their large primary branches. In the neck the deep cervical, the occipital, and the vertebral arteries send in numerous branches, in the back the intercostal arteries, and in the loins the lumbar arteries. These vessels pass in at the vertebral foramina, and send branches to the spinal membranes as well as to the bones themselves.

The blood which is returned from the spinal cord and its membranes, as well as from the vertebræ, is poured into a very intricate plexus of veins which surrounds the dura mater on its lateral and posterior surfaces, ramifying among the lobules of soft fat by which the exterior of that membrane is invested. This plexus is less intricate in the dorsal than in the cervical or lumbar regions; it communicates very freely with the plexus of veins which lies on the exterior of the vertebral laminæ and processes (the *dorsi-spinal veins* of Dupuytren). In front of the dura mater and situate between the outer edge of the posterior common ligament of the vertebræ and the pedicles, we find two remarkable venous sinuses which extend the whole length of the vertebral column, from the occipital foramen to the sacral region (*fig.* 1). These veins are loosely covered by a thin process, which is prolonged from each margin of the posterior common ligament, and is sufficiently transparent to allow them to be seen through it without removing it. They have been known since the time of Fallopius, and were described by Willis as the *longitudinal spinal sinuses*. In calibre they present many inequalities, being dilated at one part and constricted at another, according to the number and size of the vessels which communicate with them. The sinuses of opposite sides run parallel to each other and communicate by cross branches, which pass between the posterior surface of the body of each vertebra and the posterior common ligament. These cross branches present the same characters as the sinuses themselves, being of variable calibre, and presenting the greatest degree of dilatation at their middle. At this point these branches receive veins which emerge from the spongy texture of the bodies of the vertebræ (*basi-vertebral veins* of Breschet) (*fig.* 2, *d*). The vertebral

Fig. 1.

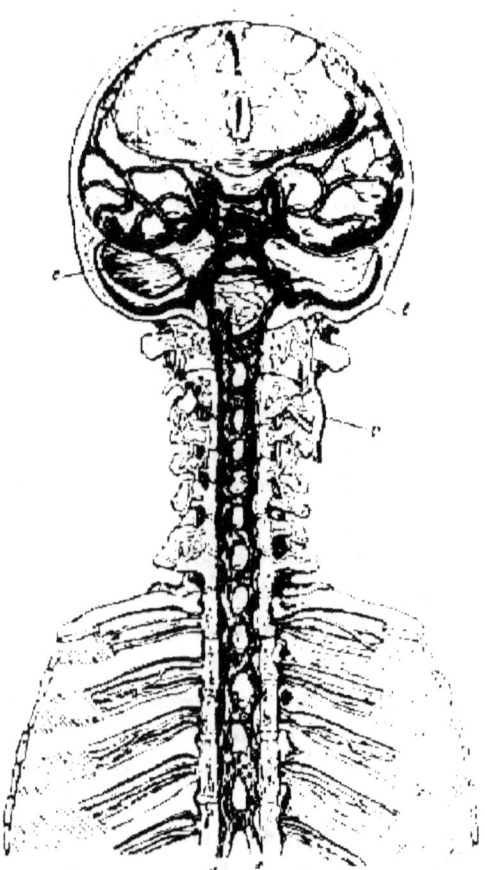

Spinal sinuses viewed from before.
(After Breschet.)

The anterior part of the basis cranii and the face have been removed, as also the bodies of the vertebræ.

l, lateral sinus descending to form its junction with the jugular vein; *c*, cavernous sinus; *v*, vertebral artery; *s*, the longitudinal sinuses with their transverse connecting veins, lying immediately behind the bodies of the vertebræ. The inferior petrosal and the cavernous sinuses appear like continuations of the spinal sinuses within the cranium, and the transverse and circular sinuses are analogous to the transverse spinal branches.

Fig. 2.

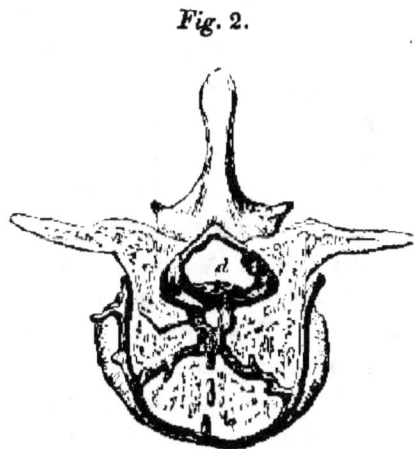

Basi-vertebral veins, converging from the spongy structure of the body of the vertebra.

sinuses diminish in size at the highest part of the vertebral canal, and passing through the anterior condyloid foramina, communicate with the internal jugular veins. In the sacral region they diminish considerably likewise, and disappear in becoming continuous with the lateral sacral veins and other small veins in that region; and they communicate with the deep and superficial vertebral veins in the neck, with the intercostal veins in the back, and with the lumbar ones in the loins. They evidently differ from the sinuses of the cranial dura mater in not being enclosed between two layers of fibrous membrane as those vessels are.

Bloodvessels of the cranial dura mater.—The bloodvessels of the cranial are much more numerous than those of the spinal dura mater, in consequence, no doubt, of the former membrane performing the office of a periosteum to the cranial bones. The arteries are derived from numerous sources; in front from the ophthalmic and ethmoidal arteries, in the middle from the internal maxillary artery by the *middle meningeal,* which enters the cranium at the foramen spinosum, and by small branches from the internal carotid which have been called *inferior meningeal* arteries. Posteriorly the vertebral, the occipital, and the ascending pharyngeal supply branches which go by the name of *posterior meningeal arteries.*

The veins of the dura mater are formed similarly to those in other parts, being derived from radicles which take their rise in the membrane itself as well as from the osseous walls of the cranium, from the diploïc veins of those bones. (See *Cyclopædia of Anatomy,* art. BONE, *figs.* 187, 188, vol. i.) All of them, with the occasional exception of one or two which accompany the middle meningeal artery and pass out at the foramen spinosum, pour their blood into the great venous canals enclosed between the laminæ of the dura mater, which are called *Sinuses.*

The sinuses of the cranial dura mater.—At certain situations, processes of the inner membrane of the venous system are included in canals formed by the separation of the laminæ of the dura mater. The channels that are thus formed for the passage of the venous blood do not admit of being dilated beyond a certain size, and in this consists an important peculiarity in the venous system within the cranium. These channels empty themselves into the internal jugular vein, which thus forms almost the sole channel by which the venous blood is returned from the brain and its membranes as well as in a great measure from the bones of the skull. And thus is explained the rapid influence which is produced upon the brain by any serious impediment to the passage of the blood through the superior vena cava.

It is important to notice that the sinuses communicate with and receive blood from certain external veins which carry blood derived from parts exterior to the cranium. Among these may be enumerated the ophthalmic vein, and several small veins in the neighbourhood of the mastoid and condyloid processes, and in the parietal bones.

The following sinuses may be described.

The superior longitudinal sinus.—This sinus corresponds to the superior margin of the falx cerebri. It commences very narrow by one or two small veins from the dura mater in the vicinity of the crista galli and cribriform plate of the ethmoid bone. Thence it proceeds backwards, gradually increasing in calibre, and it terminates a little above the internal occipital protuberance by communicating with a small cavity or reservoir, situated between the layers of the dura mater there, which is called *Torcular Herophili*. If a vertical section of this sinus be made in the transverse direction, it will be seen to be triangular in shape, the apex corresponding to the falx, the base slightly curvilinear and lodged in the groove which passes along the median line of the cranial vault. When the sinus is laid open in its length by slitting up its superior wall, we find that its sides are perforated by a great number of minute orifices, which are the openings of veins passing into it from the dura mater and from the brain itself. These veins pass into the sinus chiefly at right angles to it, or in the direction from behind forwards; a few, situate in front, enter the sinus from before backwards. In the interior of the sinus we observe little bands (*trabeculæ* of Haller, *chordæ Willisii*), stretching across from right to left, connected only with the lateral walls and leaving a free space above and below them. These bands are numerous, and various as regards breadth. Haller has seen them so numerous that they appeared like a septum dividing the sinus into two portions, of which the superior was the larger.

The walls of the sinus, towards its inferior angle, have frequently a cribriform appearance, which puts on somewhat the aspect of erectile tissue. There is no appearance of valves in the interior of the sinus; frequently, however, by the oblique entrance of a small vein into the sinus a fold is formed near the venous aperture, which, under the retrograde pressure of the column of blood, might close the orifice. When the veins open into the sinus from behind forwards, it is not improbable that they may be protected from the regurgitation of the blood by this mechanism.

Several of the small bodies, previously alluded to by the name of Pacchionian glands, project into the interior of this sinus. They appear as if they had worn their way by pressure and friction through the walls of the sinus, and it is here that the appearance of an erectile structure is most manifest. We cannot suppose that these bodies are bathed in the blood of the sinus, but rather that they push the lining membrane of the sinus before them. These bodies have been supposed to be natural structures destined to perform a mechanical office somewhat on the principle of the ball-valve, but they are frequently absent altogether, and when present they have no constant relation to the venous orifices.

The inferior longitudinal sinus (sinus inferior falcis) is a small vein lodged in the inferior part of the falx cerebri, running parallel to and a little above its inferior margin for about the two posterior thirds of its length. It gradually increases in calibre from before backwards, being formed by tributary veins from the falx; it opens into the strait sinus.

The strait sinus is situated on the middle line, at the place where the falx cerebri unites with the tentorium cerebelli. It seems to be enclosed between the layers of the former. Like the other large sinuses, it presents in its section the form of a triangle, whose base is inferior. Its direction is from before backwards and a little downwards, with a slight degree of curvature corresponding to that of the tentorium. At its commencement it corresponds to the space between the posterior reflected portion of the corpus callosum and the quadrigeminal bodies, and here it receives the inferior longitudinal sinus, and two large veins *(venæ magnæ Galeni)*, which carry the blood from the interior of the ventricles. It opens into the conflux of the sinuses or torcular by a round opening, or sometimes by two openings separated by a slip of fibrous membrane. This sinus likewise receives veins from the inferior surface of the posterior and middle lobes of the brain, and from the superior surface of the cerebellum.

At the posterior extremity of the straight sinus we find a reservoir somewhat polygonal in shape, which corresponds to the occipital protuberance; this is called the *Torcular Herophili*,* (the press of Herophilus,) the conflux of the principal sinuses of the dura mater; it has six openings, one for the superior longitudinal sinus above; one for the straight sinus in front; two for the lateral sinuses on each side; and two for the occipital sinuses inferiorly *(fig.* 3, *t*).

Lateral sinuses.—From each side of the conflux of the sinuses, there proceeds in a somewhat serpentine course outwards, downwards, and forwards, a wide canal, the largest of the sinuses, which conveys the blood from the torcular to the internal jugular vein. A groove exists on each side of the internal occipital protuberance, for the reception of this sinus, which marks the occipital bone, the mastoid portion of the temporal, and a small portion of the occipital bone again. In a great portion of their course, the lateral sinuses correspond to the posterior margin of the tentorium cerebelli, as far

* This absurd name might with great advantage be discarded, for it seems quite uncertain what precise part Herophilus intended to apply it to. The term proposed by Cruveilhier is much better, *the occipital conflux of the sinuses.* Various other names were applied to it formerly, such as *Lacuna, platea, pelvis, laguncula.*

THE CRANIAL SINUSES.

Fig. 3.

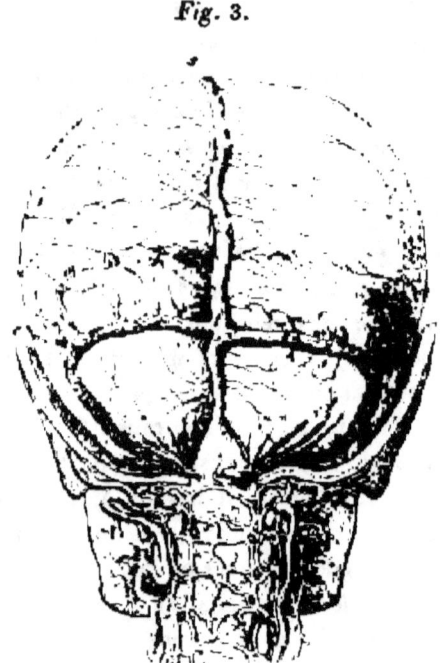

Posterior part of the cranium removed, to shew the dura mater and the superior longitudinal, and the lateral sinuses, with the torcular Herophili.

e, lateral sinus; *t*, torcular Herophili; *s*, superior longitudinal sinus.

forwards as the mastoid portion of the temporal bone. Here each sinus winds downwards to reach the jugular foramen in the posterior lacerated opening. These sinuses are never equal; that of the right side being, with few exceptions, the larger, a circumstance which Vicq d'Azyr, Söemmering, and Rudolphi attributed to the fact that most persons sleep on the right side, on which account the blood is apt to accumulate there. They are more capacious at their termination in the jugular veins than at their commencement from the torcular.

The inner surface of this sinus is like that of all the others; it is not, however, traversed by any of the bands which are found so numerous in the longitudinal sinus. Cruveilhier states that he once saw in the horizontal portion of this sinus, a few of the *Pacchionian* bodies.

In its course each lateral sinus receives veins from the inferior surface of the brain and superior of the cerebellum; it also receives the superior petrosal sinus near the base of the petrous portion of the temporal bone. A large mastoid vein communicates with this sinus and penetrates to the exterior, where it forms one of the principal sources of the occipital vein, thus establishing a free and direct communication between the circulation within and that without the cranium.[*] Near the jugular foramen the lateral sinus receives the inferior petrosal.

None of the sinuses has been more frequently the seat of inflammatory disease than the lateral. Being the principal channel for the return of the venous blood from the interior of the skull, a slight morbid action within them can scarcely fail to induce a material derangement of the cerebral circulation, and the nearness of their position to the cerebellum and to the posterior lobes of the brain renders it very unlikely that those parts would escape participating in any acute disease which might arise within it.

Occipital sinuses.—These are small veins lodged between the layers of the falx cerebelli. They collect the blood from the dura mater and from the cranial bones in the immediate vicinity of the posterior margin of the foramen magnum, and from thence they pass upwards and inwards to open into the lower part of the torcular. Cruveilhier suggests that the direction and position of the occipital sinuses are best indicated

---

[*] Cruveilhier, An. Desc. t. iii. p. 268.

by describing them as being the cords of the arcs which the lateral sinuses form.

Petrosal sinuses.—These sinuses are so named from their connection with the petrous portion of the temporal bone. The *superior* petrosal sinus corresponds, on each side, to the posterior superior edge of the petrous bone, along the three outer fourths of which a groove exists for its reception. This groove is interrupted in front by a depression in which the fifth nerve is lodged, so that at this place that nerve lies between the sinus and the bone. The superior petrosal sinus is about large enough to contain an ordinary sized surgeon's probe. It communicates with the lateral sinus posteriorly and with the cavernous sinus in front, and in its course it receives several small veins from the dura mater in the middle fossa of the cranium. It receives a vein from the anterior portion of the corresponding hemisphere of the cerebellum, and also, sometimes, one from the inferior surface of the brain. Small veins from the pons Varolii empty themselves into its anterior extremity.

The *inferior* petrosal sinuses also form an additional channel of communication between the lateral and cavernous sinuses. They are larger but shorter than the superior. In situation they correspond to the interval between the petrous bone and the occipital. They open into the inferior portion of the lateral sinus just before it unites with the jugular vein.

Transverse sinus.—This sinus establishes a communication between the petrosal and cavernous sinuses of opposite sides across the basilar process of the occipital bone. Sometimes there are two running parallel to each other. Cruveilhier states that the capacity of this sinus is much greater in old than in young subjects.

Cavernous sinuses.—In point of shape these sinuses differ considerably from all the other sinuses of the dura mater. They are venous reservoirs, situated on each side of the sella

Turcica, from which they are separated by the internal carotid arteries. Their name is derived from the spongy appearance which they present in their interior, owing to the existence of some filaments within them, which, by their interlacement with each other, form a reticular texture there. It was formerly supposed that the carotid arteries lay in the cavity of these sinuses and were bathed by their blood; but it is easy to demonstrate by a little careful dissection that the inner membrane of the sinus adheres loosely to the outer wall of the artery, and that the sixth nerve passes between them. In the outer wall of each cavernous sinus there are channels for the reception of those nerves, which pass from the cranium into the orbit. These are the third nerve which is placed highest up, the fourth nerve which holds the next place, and the ophthalmic portion of the fifth. The cavernous sinus receives at its anterior extremity the ophthalmic vein, which collects the blood from the eye-ball and other structures within the orbit, and which communicates also with the angular vein and with the frontal vein. (Hence the injected state of the vessels of the eye-ball when the brain is congested, as in fever.) Veins from the inferior surface of the anterior lobes of the brain also open into it, also some from the middle lobe and from the dura mater. Posteriorly it communicates with both the petrosal sinuses, and veins from the cranial bones open into its superior wall.

Circular sinus.—A communication is established between the cavernous sinuses by means of the *circular* or *coronary* sinus which embraces the pituitary body, one portion lying in front of it and the other behind it, both opening by a common free orifice into the right and left cavernous sinuses. The posterior portion of the circular sinus is much larger than its anterior portion. Its size is much greater, according to Cruveilhier, in old subjects than in young ones. It receives small veins from the pituitary body, and also from the sphenoid bone and from the dura mater.

It is impossible to examine this complicated arrangement of venous channels in connexion with the dura mater of the brain without admiring the beautiful provision which it affords against the undue accumulation of blood in the venous system within the cranium. In the first place, we observe that these veins do not admit of dilatation beyond a prescribed extent, by reason of their being enclosed between layers of an inelastic and inextensible membrane. Next, we remark the safety provision which is afforded by the frequent communication between them, so that if one channel were altogether closed or materially contracted, there are many others by which the blood could return. Nor is a local congestion likely to take place to any extent, for such is the freedom of communication between the sinuses and the veins of the exterior of the cranium, that (all being devoid of valves) an overflow would readily be received by the latter without the least impediment. Lastly, we learn the great importance and value of local depletion as an agent for relieving vascular fullness within the head, owing to the free communication between the extra- and the intra-cranial circulation, and especially of the veins; and we may infer from anatomy that local depletion would most probably be more serviceable than general, for although the latter would diminish the amount of the mass of circulating fluid, it would not affect the relation between the venous and arterial systems, whilst it is evident that the former must affect the venous system more directly than the arterial. Moreover, the free communication between the circulation within and that without the cranium may explain somewhat the advantage that is often derived from the application of an intense cold to the external surface of the head.

Of the pia mater. (Tunica intima vel *vasculosa.)*—The pia mater is the most internal membrane of those which have been enumerated as belonging to the spinal cord and brain.

Pia mater of the spinal cord.—This membrane stands in precisely the same relation to the spinal cord as the neurilemma does to the nerves; and as long as the spinal cord could be, as it formerly was, regarded merely as a bundle of nervous fibres, the analogy of this membrane to the nervous sheath would be perfect. It is composed almost entirely of white fibrous tissue; it closely invests the cord and supports the minute bloodvessels which carry the nutrient fluid to it. Not only does it thus form a complete sheath to the cord, but it likewise sends in processes which dip into the anterior and posterior median fissures of that organ. That which passes into the anterior median fissure is a true fold or duplicature of the pia mater; but the posterior fissure, which is much narrower than the anterior, is occupied only by a single and extremely delicate layer, which at some parts almost entirely disappears, and seems to consist merely of a few minute capillary vessels. The pia mater becomes continuous with the neurilemma of the roots of the nerves on each side of the cord, and at its inferior extremity it tapers in accordance with the shape of the spinal cord, and is prolonged as a delicate thread which is inserted into the extremity of the dura mater. This prolongation is quite gradual, so that at the upper part it encloses a portion of the medullary substance of the cord; in the greater part of its extent, however, it is merely a membranous thread, and, therefore, goes by the name *filiform prolongation* of the pia mater (*filum terminale*). The late Dr. Macartney used to regard it as highly elastic, but my friend Mr. Bowman has called my attention to the fact that it consists almost entirely of white fibrous tissue, which cannot confer elasticity. If a portion removed from the cord be stretched, it will be found to possess very little elasticity; but if the cord be held up by the filiform prolongation, and a slight jerking movement be communicated to it, it may be made to dance about as if by the elastic reaction of the filiform process. The movement which may be thus produced is very well calculated to deceive, and Dr. Macartney must have founded his opinion upon that experiment alone, omitting to try the effect of stretching a detached portion of the process. The fact is that when the cord is suspended in this way, the pia mater becomes stretched, and its anterior and posterior portions are approximated and the cord flattened; when it is raised with a jerk, this tension of the pia mater is diminished, and the cord returns to its previous form until it falls again, stretches the pia mater, and becomes once more flattened, producing a degree of reaction which favours its elevation, but which alone would be insufficient for that purpose. Thus it appears that the elastic reaction, which Dr. Macartney attributed to the filiform process, is in reality due to the compression and consequent flattening of the cord by the tension of the pia mater. It should be stated, further, that this process is not formed of pia mater alone, but also of a continuation of the ligamentum denticulatum on each side, to be described by-and-bye.

The pia mater is abundantly supplied by bloodvessels, many of which are extremely tortuous. These vessels are derived from the anterior and posterior spinal arteries. Along the anterior surface of the spinal cord in front of the anterior median fissure there is a narrow band of fibrous tissue which is stretched across this fissure like a bridge, and occupies its whole length. No such arrangement exists on the posterior surface.

The pia mater of the spinal cord possesses considerable strength and density. The nervous matter may by pressure be squeezed out of it, leaving a hollow cylindrical membrane, or it may be dissolved out by the action of liquor potassæ. In the quite recent state, while the cord is as yet firm, the pia mater may be readily dissected off, its adhesion to the cord being through the medium of numerous exceedingly minute capillary vessels. On its exterior the pia mater adheres to the visceral layer of the arachnoid membrane by means of a loose fibrous tissue.

Pia mater of the brain.—In tracing the pia mater of the spinal cord upwards, it will be found gradually to become much thinner and more delicate as it passes from the medulla oblongata to the hemispheres of the cerebellum and cerebrum. In connexion with these latter parts it becomes of extreme tenuity, and owes its physical tenacity chiefly to the intimate connexion of the visceral layer of the arachnoid membrane with it. The cerebral pia mater is almost exclusively composed of numerous ramifications of minute vessels which are accompanied by white fibrous tissue in small quantity. These vessels divide and subdivide to the last degree of minuteness, and are admirable objects for examining the structure of capillary vessels. The pia mater adheres closely to the whole surface of the brain, cerebellum, and connecting parts, and numberless vessels pass from it into the nervous substance in contact with it. On the surface of the brain it adheres to the superficial grey matter and sinks down into the sulci or furrows between the convolutions. Wherever there is a depression or fissure of the brain, the pia mater is found dipping into it. It likewise sinks into the fissures between the laminæ of the cerebellum.

We shall obtain, however, a very inadequate notion of the extent of the pia mater, if we confine our examination of it to the exterior of the brain and cerebellum. At certain situations this membrane is continued into the cavities or ventricles of these organs, where it doubtless fulfils some office connected with the support and nutrition of certain parts of them. These situations are four in number, as follow: on each side, the fissure between the crus cerebri and the middle lobe of the brain, behind, the transverse fissure between the cerebellum and cerebrum, and, lastly, the inferior extremity of the fourth ventricle.

Choroid plexuses of the lateral ventricles.—These are apparently folded processes of the pia mater which enter the inferior part of the lateral ventricles on each side, and are continued upwards and forwards to the middle portions of those cavities, where they become continuous with each other in the foramen commune anterius, and with a middle process, the velum. Each choroid plexus forms a somewhat cylindrical process, which, when traced from below upwards and from behind forwards, will be found to follow the direction of the lateral ventricle as far forwards as the apex of the horizontal portion of the fornix, gradually diminishing in thickness, and assuming the character of a simple membranous expansion. It projects freely into the cavity of the ventricle, having no connection with its walls, excepting along the margins of the fissure, at which it enters, where the membrane of the ventricle adheres to it, being probably reflected upon it.

Very numerous and tortuous bloodvessels are contained in these

Fig. 4.

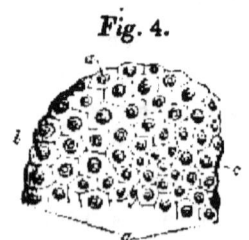

Choroid plexus of lateral ventricle in the Sheep, shewing a villous process, highly magnified, and the epithelium.
(After Valentin.)
a, villus; b, epithelium; c, nucleus of epithelium.

processes, forming a plexus which has given name to the folds themselves. The surface of each choroid plexus presents many slight projections or folds resembling villi, in which are contained loops and plexiform anastomoses of minute vessels, very similar to the arrangement of the vessels of the villous processes of the chorion of the ovum, or those of the tufts of the placenta. These vessels are surrounded by an epithelium which has much the appearance of that of serous membranes. From the great number of these vessels and from the delicate nature of the epithelial covering which surrounds them, it is plain that the choroid plexuses are well suited either for the purpose of pouring out fluid or of absorbing it.

Fig. 5.

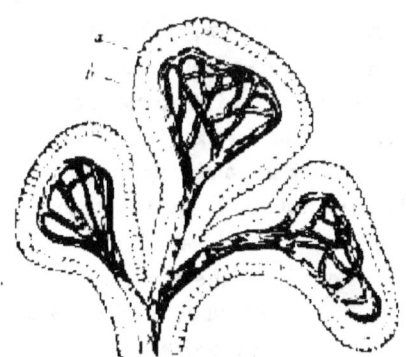

Side view of villi of the choroid plexus of the lateral ventricle in the brain of a Goose, to show the disposition of the bloodvessels. Not to obscure the view of the bloodvessels, the edge of the epithelium only has been shown.

(*After Valentin.*)

a, epithelium; *b*, bloodvessels.

The epithelium may be best seen by examining the edge of a fold. It becomes very distinct when acted upon by acetic acid. As its particles are very delicate and consist only of a single layer, they are easily detached. The cells of epithelium are most of them six-sided, and contain a clear nucleus, or several minute granules. Valentin states that cilia may be seen playing upon this surface, especially in the embryo. I have observed the peculiar punctiform or spiriform formations to which he alludes, which look like the remains of former vibratile cilia.

Velum interpositum.—(*Toile Choroidienne*, Vicq d'Azyr.)—The choroid plexuses are connected to each other by the *velum interpositum*, which is a triangular fold of pia mater that passes in at the transverse fissure between the upper surface of the tubercula quadrigemina and the posterior reflected portion of the corpus callosum. This process is continuous with the pia mater of the inferior surface of the posterior lobes of the brain, and with that of the superior surface of the cerebellum, and it therefore consists of two laminæ; as it passes forwards, it sends downwards a little process which embraces the pineal body; it forms the roof of the third ventricle, being interposed between that cavity and the fornix, (hence its name,) and at its sides as well as its apex its continuity with the choroid plexuses may be readily demonstrated. At its anterior extremity it corresponds to the foramen commune anterius. The velum interpositum is best exposed in the dissection from above downwards by removing carefully in succession the corpus callosum and the fornix. In raising the velum itself, in order to disclose the cavity of the third ventricle, it is necessary to be very careful, as from the intimate connexion which the pineal body has with it towards its base, that body may be readily disturbed from its position.

Choroid plexuses of the fourth ventricle.—The choroid plexuses of the fourth ventricle are two small processes of pia mater united along the median line, presenting the same villous character as those of the lateral ventricles. These folds seem as if they had been pushed up into the fourth ventricle by the lower laminæ of the inferior vermiform process. Their position may be best seen by opening the fourth ventricle from above, where they will be found lying on each side of that portion of the median lobe of the cerebellum which stops up the inferior extremity of the fourth ventricle. These plexuses are in every respect similar, as far as regards structure, to the larger ones which are found in the lateral ventricles, and, like them, exhibit a delicate epithelium upon their surface. Upon the centre of each epithelium cell Valentin states that a pigment corpuscule is deposited. (*Fig.* 6.)

Fig. 6.

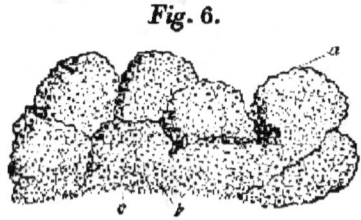

A highly magnified villus of the choroid plexus of human cerebellum. (*After Valentin.*)

a, the villus; b, the epithelium cells; c, the nuclei.

These internal processes of the pia mater contain minute crystalline formations, a kind of very fine sand, which, however, is not constantly present.

The grains are deposited in the meshes of the vascular plexuses. Sometimes they accumulate in masses so as to be visible to the naked eye or easily recognized by the touch. In general, however, they are microscopic, in form globular, and connect themselves with the minute vascular ramifications like little bunches of grapes. They are found principally in the choroid plexuses of the lateral ventricles, and in that portion of the velum interpositum which embraces the pineal body. In the former they are most numerous at that part which was called by the Wenzels *glomus*, where the choroid plexus turns up from the inferior cornu into the horizontal portion of the lateral ventricle.* As regards chemical composition this sabulous matter consists chiefly of phosphate of lime with a small proportion of phosphate of magnesia, a trace of carbonate of lime, and a small quantity of animal matter.

The pia mater adheres very closely to the surface of the brain, coming for the most part into contact with the grey or vesicular matter. When a portion of it is carefully raised in a fresh brain, numberless extremely minute bloodvessels are seen passing from it into the cerebral substance. These are the principal nutrient vessels of the brain. On its outside the pia mater adheres partially to the arachnoid membrane. At those points which correspond to the convex portions of the convolutions the adhesion of the arachnoid to the pia mater is close; but at other places the latter membrane separates completely from the former.

The pia mater of the brain differs from that of the spinal cord in its great delicacy and tenuity; it wants the strength and density of the latter membrane. This is owing to its being composed almost entirely of extremely minute and delicate bloodvessels, whilst the spinal membrane consists chiefly of white fibrous tissue. The bloodvessels of the former are infinitely more numerous than those of the latter, and the reason of this probably is that the cerebral membrane is chiefly in contact with grey matter, which requires a great quantity of blood, but the spinal membrane immediately embraces white matter, which is much less vascular.

It is important, in a pathological point of view, to notice

* See Van Ghert de plexubus choroideis, Utrecht. 1837; Valentin, in Soemmering Anat., and Bergmann, über die innern Organisation des Gehirns. The last author states that he has seen the sandy deposit excessive in connexion with mental derangement.

that this membrane is the medium of nutrition, not merely to the nervous matter of the brain and cord, but also to the arachnoid membrane which is immediately adherent to it, and to which it bears the same relative position as the sub-serous areolar tissues elsewhere to their respective serous membranes. Hence the difficulty, if not the impossibility, of adopting distinctions which systematic writers endeavour to make out between arachnitis and superficial inflammation of the brain. It is physically impossible that there shall be arachnitis without serious disturbance of an inflammatory kind in the circulation of the pia mater, nor can this exist without affecting the superficial layers of the grey matter of the convolutions. It may, therefore, be confidently affirmed that arachnitis, when affecting that portion of the arachnoid membrane which covers the hemispheres of the brain, is synonymous with inflammation of the superficial layers of the grey matter of the convolutions. Whatever be the point of departure, it seems impossible that inflammation of the one can exist without a similar and equal affection of the other. And thus we may explain the apparently anomalous statement of authors that inflammation of the arachnoid should give rise to a more violent train of symptoms than deep-seated inflammation of the brain. The real difference is, not between membranous and cerebral inflammation, but between an inflammatory affection of the superficial grey matter of the convolutions, the great source and seat of the physiological activity of the brain, and a similar morbid action of the more central white substance, the function of which is in a certain sense subservient to that of the superficial grey matter.

Of the arachnoid membrane.—This membrane is intermediate to those already described. We have preferred giving the description of it last, because to understand it demands an acquaintance with the details of both those membranes.

The arachnoid is a great serous membrane pervading the entire cranio-spinal cavity. Its parietal layer adheres intimately and inseparably to the inner surface of the dura mater both cranial and spinal, and its visceral layer is attached to the outer surface of the pia mater.

In point of structure and general disposition the arachnoid membrane resembles other serous membranes, so much as to render it inexpedient to enter into any minute comparison of them. It will only be necessary to refer to such peculiarities of arrangement as may arise from the anatomical characters of the nervous centres with which it is connected.

Spinal arachnoid.—The serous character of the spinal arachnoid is best seen by examining a transverse section of the spinal cord and its membranes. If the section be made across the interval between two sets of spinal nerves, the visceral and parietal layers of the membrane may be seen in contact with each other; the parietal layer closely attached to the dura mater, the visceral layer adherent to the pia mater of the spinal cord so loosely as to leave a considerable space between it and the outer surface of that membrane.

Fig. 7.

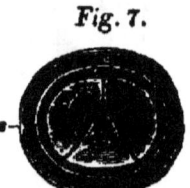

Transverse section of spinal cord and its membranes between the fifth and sixth cervical nerves.
(After Arnold.)
v, visceral layer of arachnoid membrane; s, sub-arachnoid space; c, arachnoid cavity.

We may here notice an important distinction which the student of this portion of anatomy will do well to note particularly, namely, that the space between the two layers of arachnoid membrane is the *arachnoid bag or sac*, in which it is very rare for any fluid to accumulate; and that that between the visceral layer of the arachnoid and the pia mater is the *subarachnoid cavity*, in which, as will be shown by-and-bye, a considerable quantity of fluid exists in the natural state.

When the section is made on a level with the nerves as they emerge through the dura mater, we may notice the manner in which the arachnoid membrane is prolonged upon them in the form of a loose sheath, forming little culs-de-sac at the orifices through which they escape.

Fig. 8.

Transverse section of the same on a level with the fifth cervical nerves.
(After Arnold.)

The same parts are displayed as in the last figure, and the reflection of the arachnoid at the exit of the nerves is seen.

af, anterior fissure ; *n, n,* spinal nerves.

In the interval between each pair of nerves, we find a triangular process of fibrous membrane which is inserted by its apex into the dura mater. This process lies in the subarachnoid cavity and adheres by its base to the pia mater. It seems to pierce both layers of the arachnoid, or to pin them down, as it were, to the dura mater.

At the foramen magnum the spinal arachnoid may be seen to be continuous with that of the brain, and here its visceral layer invests the medulla oblongata loosely. Inferiorly we trace the membrane down quite to the lowest extremity of the dura mater, and in this region the visceral layer is particularly loose and free, as it lies over the cauda equina.

When the dura mater is carefully slit up along either the anterior or the posterior surface, the arachnoid sac is laid open. It does not always happen that the parietal layer separates very readily from the visceral: frequently the two layers adhere firmly at several minute points, yet this adhesion is effected without any connecting membrane, and appears to arise from the two layers becoming dried at several corresponding points, and thus being, as it were, glued together. We may frequently observe this in specimens that have been some time kept in spirits. This point is deserving of notice, as these adhesions might be (and indeed they have been) noted as of a morbid nature.

The visceral layer of the spinal arachnoid is connected to the pia mater by means of a number of long filaments of fibrous tissue which interlace slightly, and in the areolæ thus formed the fluid is contained. This tissue is most distinct and abundant in the cervical region, and exists in very small quantity in the dorsal. It ceases nearly altogether over the cauda equina. Numerous minute bloodvessels are also to be found in it passing from the pia mater to the arachnoid. Majendie gives to this tissue the name "*tissu cellulo-vasculaire sub-arachnoide.*"

In general the adhesion of the visceral layer of the arachnoid to the subjacent pia mater is closer along the posterior than along the anterior surface of the cord.

Along the posterior surface of the cord on the median line, the sub-arachnoid space is divided by means of a septum, which is most perfect in the dorsal region, but which in the lumbar and cervical regions is cribriform or pectiniform, as may be shown by pouring quicksilver on either side of it, which will be retained in the dorsal region, but will readily pass from right to left in the other situations. It is highly probable that this septum is a modified portion of the sub-arachnoid tissue.

The existence of this septum (erroneously described as complete) dividing the posterior part of the sub-arachnoid space into a right and a left portion, appears to have led to the opinion that this space is lined by another serous membrane, which has been called *the internal arachnoid*, by which the fluid is supposed to be secreted, and that the septum is formed by the reflection of its visceral into its parietal layer along the median plane. But there are many objections to this hypothesis. In the first place, if the septum were formed by the reflection of a serous membrane, it would be complete, and not a very imperfect one such as it is; it ought to resemble the mediastinum in the chest, or one of the processes of the peritoneum in the abdomen. Secondly, it is quite contrary to all experience to find the cavity of a serous membrane in the normal state traversed by a quantity of filamentous tissue, as the sub-arachnoid space is throughout a great part of its extent. Thirdly, were there a serous membrane in this space, the microscope ought to detect an epithelium on its inner surface, but such a structure does not exist here. Lastly, such a serous membrane must necessarily be continued into the encephalic sub-arachnoid space. But the close adhesion of the visceral layer of the arachnoid to the pia mater, opposite to the prominent parts of the cerebral convolutions, seems quite incompatible with such an arrangement.

Cerebral arachnoid.—The cerebral portion of the arachnoid exhibits essentially the same general arrangement as the spinal portion. Its parietal layer adheres very intimately to the pia mater at certain points, leaving in the intervals a considerable space for the accumulation of liquid. If we trace it over the surface of the hemispheres, it will be found to give them that smooth and uniform character which is always distinct on the recent healthy brain. The arachnoid passes from convolution to convolution, adhering closely to the pia mater over the most convex portions of those convolutions, but allowing that membrane to separate from it in the intervals between them, and to dip down to the bottom of the sulci. Hence liquid accumulated in the cerebral sub-arachnoid space will be found to take the direction of the intergyral sulci, and to cause the membrane to bulge opposite to them; and if air be blown underneath the arachnoid, it will be found to take the tortuous course of these sulci.

The arachnoid sinks into the great longitudinal fissure of the brain, lining the surfaces which bound it on each side, and passing across from right to left beneath the inferior margin of the falx, and above the corpus callosum.

On the base of the brain, the arachnoid has the same arrangement on those parts where there are convolutions, as on the superior and lateral surfaces of the hemispheres. It passes over the fissure of Sylvius from the anterior to the middle lobe, and here its distinctness from the pia mater may be clearly demonstrated; here too it appears much stronger and more opaque than elsewhere, which is probably due to the existence of an increased quantity of fibrous tissue beneath it.

In that space on the base of the brain which is bounded on each side by the middle lobes, and which is limited in front by the optic nerves and behind by the pons Varolii, the arachnoid membrane stretches across from one middle lobe to the other, leaving a considerable vacancy between the tuber cinereum and the pons, in which it is connected to the pia mater by several long filaments similar to those which are met with on the surface of the spinal cord. This space is favourable for the accumulation of fluid, and it communicates in front with the fissures of Sylvius and other deep fissures into which fluid might make its way. Cruveilhier calls it the *anterior sub-arachnoid space,* and regards it as the principal reservoir of the cranial serosity. Immediately in front of it we observe that the arachnoid membrane is continued around the infundibulum to the pituitary body.

In tracing the arachnoid backwards from the great longitudinal fissure of the brain, we observe that it stretches down from the posterior edge of the corpus callosum to the superior surface of the cerebellum, crossing over the tubercula quadrigemina. At this situation the arachnoid is reflected upon the venæ magnæ Galeni as they pass to the straight sinus. It was at this place that Bichat described the canal which goes by his name, through which, as he thought, a process of the arachnoid membrane was carried in to line the interior of the ventricles.

The arachnoid covers the superior surface of the cerebellum and also its inferior surface, stretching across the longitudinal fissure from one hemisphere to the other, and it is also extended downwards, and a little forwards from the superior surface of the cerebellum to the posterior surface of the medulla oblongata, below the inferior extremity of the fourth ventricle. A considerable space is thus left, situate posteriorly between the cerebellar hemispheres, and bounded in front and inferiorly by the medulla oblongata, which also forms a considerable reservoir for cerebral fluid, and communicates freely with the sub-arachnoid space of the spinal canal; but as the arachnoid is tied down somewhat more closely over the posterior surface of the spinal cord, there is an appearance of constriction where the cerebral passes into the spinal arachnoid. This space is called by Cruveilhier the *posterior sub-arachnoid space (posterior conflux* of Majendie). It communicates with the anterior sub-arachnoid space through the furrows around the crura cerebelli.

Of the cerebro-spinal fluid.—In examining such a dissection of the membranes of the spinal cord as that above described, we shall find that at various points the visceral layer of the arachnoid membrane appears raised up by fluid or by a bubble or two of air from the subjacent viscus. If a small portion of this layer be taken up in a forceps, and a blowpipe be introduced into it, air may be blown underneath it, raising it up all around the spinal cord to a considerable distance from that organ. The inflation is more easily effected in the cervical and in the lumbar regions than in the dorsal, and the air will pass down quite to the lowest part of the canal of the dura mater, where the connexion of the arachnoid membrane to the cauda equina is particularly loose. In the same way coloured fluid, or some material which may assume the solid form, as size, tallow, &c. may be injected to demonstrate this anatomical arrangement. If now we examine a transverse section, it will be observed that a considerable interval exists between the visceral layer of the arachnoid and the pia mater of the cord, and that this interval is much greater in the neck and in the loins than in the back. We observe too that the spinal cord is by no means of sufficient size to fill the spinal canal, and that as a considerable interval exists between its surface and the visceral layer of the arachnoid, so also a still greater one is found between it and the inner surface of the dura mater. Now as it is of the very nature of a serous membrane that its free and smooth surfaces should always be in contact (for it is in that way that it favours the movements of the viscus with which it is connected), it is plain that the sub-arachnoid space in the spine must, during life, be kept in a state of distension, otherwise the object of a serous membrane would not be attained.

Moreover, in tracing the arachnoid membrane upwards over the medulla oblongata and the other parts of the encephalon, we observe an evident continuity between the spinal and the cranial sub-arachnoid cavity, which is most evident at the base of the brain, where the latter possesses the greatest dimensions, so that air or fluid may be readily made to pass from one to the other. This is most conspicuous in old subjects, in which the brain being small and more or less shrunken, leaves a considerable interval between its surface and the visceral layer of the arachnoid.

On opening the spinal canal in a body recently dead, the visceral layer of the arachnoid will almost always be found raised by fluid. When a portion of the posterior wall of the spinal canal is removed in a living animal, or in one just killed, the dura mater is found to be quite tense from the fluid which is accumulated within it. In a horse, whose spinal canal I opened in the dorsal region immediately after he had been knocked down in the knacker's yard, I found the dura mater perfectly tense, and semi-transparent from being stretched so firmly over fluid. Upon making a minute puncture in it, a fine stream of clear fluid was ejected with much force to a considerable distance, and immediately the dura mater became quite flaccid. By a little careful dissection through the dura mater and parietal layer of the arachnoid, it may be shewn that this fluid is not contained in the arachnoid sac, but in the sub-arachnoid cavity.

We can thus demonstrate the existence of a fluid, which during life and in a state of health occupies the sub-arachnoid cavity and maintains the two layers of the arachnoid membrane in contact with each other. This fluid is designated by Majendie the *cerebro-spinal* fluid.

The first distinct recognition of this fluid in its proper locality is due to Cotunnius, who stated the results of his observations in his memoir " de Ischiade Nervosâ," preserved in Sandifort's collection of dissertations. Cotunnius was led to the discovery by remarking the great disproportion between

40 THE CEREBRO-SPINAL FLUID.

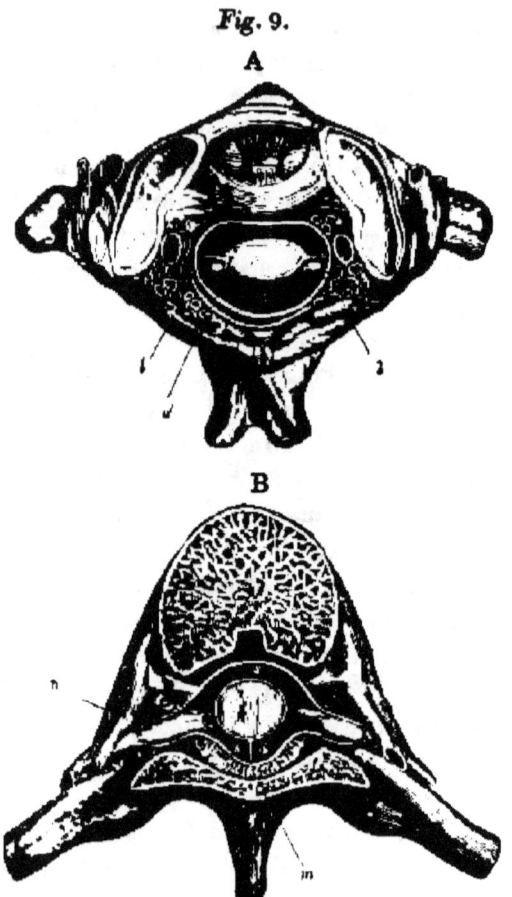

Fig. 9.

A, *Transverse section of the spine at the situation of atloido-occipital articulation.*
(*After Majendie.*)
 c, the spinal cord.
 d, the dura mater and arachnoid membrane.
 s, s, the sub-arachnoid space, divided into an anterior and posterior portion by
 l, the ligamentum denticulatum.

B, *Section in the dorsal region.*
 The same letters indicate similar parts as in A.
 m, the posterior median septum.
 n, the roots of the nerves.

THE CEREBRO-SPINAL FLUID. 41

Fig. 10.

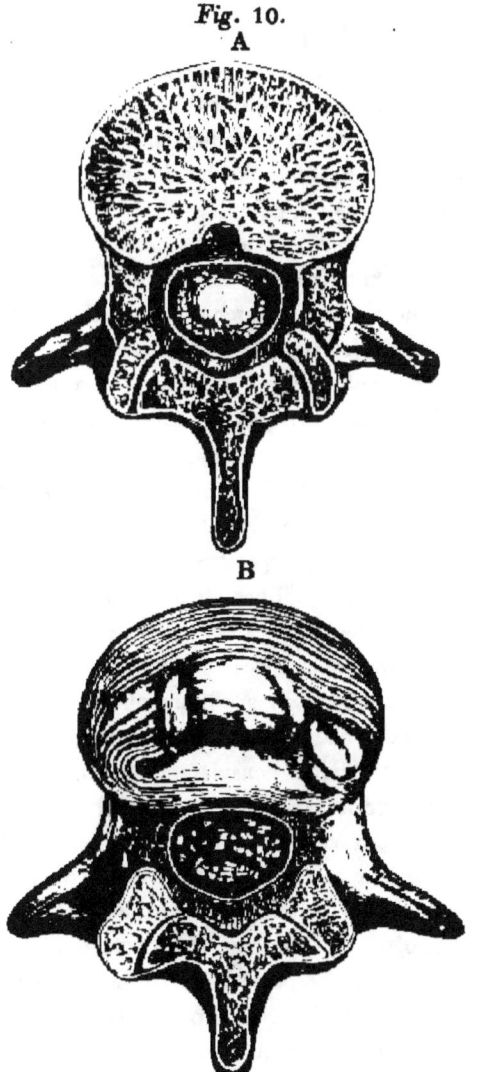

Sections of the spine in the lumbar region.
A, shews the section of the cord as well as of many roots of nerves descending to form the cauda equina.
B, shews the section of the cauda equina.
In both these regions the sub-arachnoid space is large and uninterrupted by bands or septa. The fluid permeates between and surrounds the roots of the nerves.

the dimensions of the spinal canal and the bulk of its contents, so that a considerable interval exists between the internal surface of the former and the spinal cord; and he attributes its having been so completely overlooked by previous anatomists to the fashion of opening the head before the spine, which favoured the escape of the fluid.

It is, however, to M. Majendie that we are chiefly indebted for our present knowledge of the physiological history of this fluid. Majendie's first researches were given to the public in his Journal de Physiologie for the year 1827, and he has lately collected the results of his inquiries in a volume entitled " Recherches Physiologiques et Cliniques sur le Liquide Cephalo-rachidien," and published in 1842.

The cerebro-spinal fluid is found wherever pia mater exists in connexion with brain or spinal cord, whether on the surface of these organs, or in the ventricles of the former. It serves to fill up various inequalities in the cranial or spinal walls, and it accumulates in greatest quantity in those situations where the sub-arachnoid space affords the greatest capacity.

Majendie describes four situations at which this fluid accumulates in greater quantity than at other places on the surface of the brain. The most considerable of these, which he designates the *posterior conflux*, is situated below and behind the cerebellum; it corresponds to the posterior surface of the medulla oblongata, and is covered behind by the layer of arachnoid which extends between the medulla and the cerebellum. (Vid. supr. p. 37.) It is here that, according to Majendie, a communication takes place between the fluid on the exterior and that in the ventricles, at a point corresponding to the inferior extremity of the fourth ventricle. A second, or *inferior conflux* is found immediately in front of the pons Varolii; it is situated between the crura cerebri, and contains the basilar artery. It is, in fact, only the posterior part of what Majendie designates the *anterior* conflux, which extends forwards to the commissure of the optic nerves, occupying the central depression between the middle lobes of opposite sides, and bathing in its fluid the commissure, the tuber cinereum, the infundibulum, and the trunks of the anterior cerebral arteries. It communicates with the posterior fissure beneath the crura cerebelli. The position and the extent of this conflux is indicated by the separation of the visceral layer of the arachnoid membrane over the central part of the base of the brain. Doubtless the accumulation of fluid around so many parts of important function and delicate structure, is a valuable safeguard to them against the communication of shocks from the walls of the cranium. A fourth conflux is called *superior;* it is situated behind and a little below the level of the corpus callosum, behind the pineal gland, and above the tubercula quadrigemina. It communicates around the crura cerebri with the anterior conflux, and with the posterior conflux by the fissures which separate the superior vermiform process from the hemispheres of the cerebellum. The fluid contained in it bathes the pineal gland, the tubercula quadrigemina, the superior vermiform process, and the venæ Galeni as they empty themselves into the straight sinus.

As the fluid is in contact with pia mater, it is plain that it must surround and support the roots of the nerves which proceed from both the brain and spinal cord, and that the bloodvessels which penetrate or emerge from those organs, or which ramify in the pia mater, must also be bathed by it. The fluid surrounds the nerves as they emerge from the cranium or spine, and maintains contact between the layers of arachnoid membrane which compose the sheaths that accompany them in their passage outwards. Majendie states that this fluid accompanies the roots of the fifth pair of nerves as far as the Gasserian ganglion, and that it bathes and mingles with the fibres of the ganglion itself, as well as of the three nerves which originate from it. This, however, I think extremely doubtful.

That fluid exists in the ventricles of the brain has long been known to anatomists; and it seems more probable that this fluid is secreted by the processes of pia mater which are found in all these cavities; or possibly by the membrane which lines their surface. Does the internal fluid communicate with that in the sub-arachnoid space? Majendie affirms that a communication takes place by means of an opening which is situated at the inferior extremity of the fourth ventricle. I have not been able to satisfy myself of the existence of such an opening; the following is Majendie's description of it: "The true orifice, constant and normal, by which the cerebro-spinal fluid continually passes, either to enter the ventricles or to issue from them, may be seen at the *inferior termination of the fourth ventricle*, at the place named 'le bec de la plume' by the old anatomists.

"To demonstrate the existence of this orifice it is necessary to raise up, and to separate slightly from one another, the lobules of the inferior vermiform process of the cerebellum, and without breaking any of the vascular adhesions which unite this part of the cerebellum with the spinal pia mater, we perceive the angular excavation which terminates the fourth ventricle. Its surface is smooth, even *(polie)*, and is prolonged as far as the ventricle of the cerebellum. Such is the anterior part of the orifice: the lateral and superior parts are formed by the choroid plexuses of the organ and by a horny medullary lamella (valve of Tarin), the extent of which is variable, and which adheres to the prominent border of the fourth ventricle. The form and dimensions of the opening vary with the individual, and with the quantity of cerebro-spinal fluid, so that when the latter exists in considerable quantity the opening can admit the extremity of a finger. Most frequently, when the quantity of the liquid is normal, the orifice does not exceed two or three lines in diameter in every direction, but it is frequently subdivided by vessels which pass from the medulla oblongata to the cerebellum. Sometimes the orifice is restricted by one or by both of the posteror cerebellar arteries which pass across it."

Such is the description of the orifice to which Majendie has given the high-sounding title " *Orifice des cavités encephaliques.*" He states that when fluid is injected into the spinal sub-arachnoid cavity, it makes its way into the ventricles of the brain through this orifice, a statement sufficiently difficult to prove. Cruveilhier, who seems to lean towards Majendie's opinion, admits nevertheless several weighty objections to it. The most important of these appears to me to be that the margins of the orifice which is brought into view by the method directed by Majendie, are irregular, and have the appearance of lacerated membrane. And it is recorded by M. Martin St. Ange, on the authority of Cruveilhier, that in fifteen subjects in which the latter anatomist found this orifice, its margins had the torn appearance in every one; " that around the opening, here and there, there existed the debris of membranes."*

My own opinion is that this orifice does not exist naturally, but that it is produced by the violence to which the brain is subject in its removal, or in the manipulations necessary for demonstrating it. It appears to me that the fourth ventricle is closed in the same way as the inferior horn of the lateral ven-

* Martin St. Ange. Sur les membranes du cerveau et de la moelle epiniere.

tricles, namely, by the reflection of its proper membrane from its floor on to the adjacent pia mater. This membrane is so extremely delicate that the slightest traction upon it is sufficient to disturb its connexions. Its existence may be best proved by the resistance which a probe pushed into the fourth ventricle from above experiences at its inferior extremity, a resistance, however, which a little force can overcome. Or, if the fourth ventricle be opened from the side, by a vertical section of the median lobe of the cerebellum a little to one side of the median plane, and if this be done on a brain previous to its removal from the body, or on one which has been removed with great caution, so as to occasion the least possible disturbance to the parts, it will be found that the ventricle is closed below by the reflection of its proper membrane upon the pia mater. There can be no doubt that fluid driven against this membrane with force, whether from within the ventricle or from the sub-arachnoid space, would easily rupture it.*

It is plain that if there be a direct communication between the fluid in the ventricles and that in the sub-arachnoid cavity at the inferior extremity of the fourth ventricle, it must take place through an opening in that portion of the pia mater which ascends into the fourth ventricle to form the choroid plexus. But it is not necessary to have recourse to such a supposition to account for the transmissibility of fluid from one cavity to the other, for the pia mater is evidently hygrometric, and will readily admit of the passage of fluid through it by endosmose, and it is highly probable that, if any interchange of fluid takes place between the intra-ventricular cavity and the sub-arachnoid space, it is accomplished through the influence of endosmose and exosmose, effected not merely by

* See the description of the fourth ventricle further on.

the pia mater at the inferior extremity of the fourth ventricle, but likewise by that at the inferior cornua of the lateral ventricles, and perhaps also by that of the third ventricle, at the velum interpositum. And it is worthy of remark, as tending to confirm this opinion, (which, so far as I am aware, has not previously been suggested,) that at each of these situations there is a conflux (to use Majendie's phrase) of the sub-arachnoid fluid.

Cruveilhier lays some stress upon the fact that in apoplexy the blood escapes from the ventricle into the sub-arachnoid space. For my own part, I would say that this occurrence takes place as often, if not more frequently, at the inferior cornua of the lateral ventricles, as at the fourth ventricle. And therefore, if such a fact be used as an argument in favour of the direct communication of the latter with the sub-arachnoid space, it ought equally to lead to the supposition of the existence of similar orifices at the former situations, the absence of which may be easily proved. Moreover it may be stated that blood sometimes extravasates into the arachnoid sac, breaking through the arachnoid membrane; it is, therefore, less difficult to conceive its bursting the pia mater, which is evidently more porous, and is the seat of those vessels, a morbid condition of which is the frequent precursor and the cause of the apoplectic attack.

The best way of obtaining the sub-arachnoid fluid with a view to form an estimate of its quantity, is to open the dura mater and arachnoid in the lumbar region of the spine, having previously, by means of a trephine, made a small perforation in the skull, so as to allow the pressure of the atmosphere to bear upon the cranial contents. " If," says Cotunnius, " you open the vertebræ of the loins before the head is touched, and cut the enclosed tube of the dura mater, a great quantity of water will burst out, and after all this spontaneous flux of water is spent, if you lift up the head, and shake it toward the aperture, a more plentiful stream will burst out, as if a new fountain was unlocked. In these experiments, which I made on the bodies of near twenty adults, and which I repeated at different times, I could draw off freely from the hollow of the spine four and sometimes even five ounces of water: I commonly found it very clear in such subjects, although it sometimes inclined a little to a yellow colour; but in fœtuses strangled in difficult labour, little as it was, I observed it to be always red and opaque."*

The estimate of the quantity of sub-arachnoid fluid here assigned by Cotunnius exceeds that which Majendie deduces from his experiments, who states that in general in a subject of adult age and mean size, and in moderate condition, two ounces may be regarded as the minimum quantity. Much depends upon the age and size of the subject and the state of nutrition of the nervous centres. In children the quantity is very small; in old age, when the brain and spinal cord have shrunk considerably, the quantity is large. In examining the bodies of the aged poor, as Majendie remarks, eight, ten, or twelve ounces of fluid may be obtained from the cranio-spinal cavity, according as there is a greater or less degree of atrophy of the brain.

In judging of the quantity of fluid around as well as within the cerebro-spinal centres, the time which has elapsed since death must be taken into account. As advancing decomposition favours the transudation of fluids through the tissues, it is plain that the longer this period is, the less liquid will be found; and the earlier after death the investigation takes place, the nearer will be the resemblance of the parts to their con-

* From an English translation of Cotunnius's essay, entitled, A Treatise on the Nervous Sciatica, or Nervous Hip Gout, translated by Henry Crantz. London, 1775.

dition during life. On the other hand, a very advanced stage of decomposition will favour the developement of liquid, wherever space may be found for its accumulation. It is, therefore, in vain for the pathologist to attempt to form an opinion respecting the quantity of the fluid found in the cranio-spinal cavity, unless the inspection have been made at an early period after death.

Practical men are too much in the habit of attributing morbid phenomena of the nervous system to the influence of the pressure of a liquid effusion upon the brain or spinal cord. Many facts tend to shew that in a large proportion of cases, especially in the adult, the occurrence of an increased quantity of fluid, either around those centres or within the ventricles, is *a result*, and that it is probably a result *of a conservative kind*, consequent upon a morbid change which depresses the general nutrition of those organs themselves. We have seen how the universal decay of the tissues, which characterizes old age, favours the increase of the cranio-spinal liquid, when it affects the brain and spinal cord. In examining the bodies of habitual drunkards, patients who die of delirium tremens, or of cirrhose of the liver, the quantity of fluid is always found to be considerable and the brain shrunk. In bed-ridden persons who have ceased to exercise their faculties for some time, whether for mental or bodily exertion, the same phenomena are witnessed. When there has been much anæmia, as in cases where death has terminated a protracted illness, in phthisis for example, or in persons who have died of hæmorrhage, or after excessive venesection, the nervous centres will be found to be small and the liquid in large quantity. In extreme cases of lead cachexy, in which the nutrition of the nervous and muscular tissues is materially diminished, I have observed similar appearances. And, when any partial atrophy of either the brain or the spinal cord has occurred, there will invariably be found, at a point corresponding to it on the exterior of the organ, a local accumulation of fluid occupying a depression on its surface which has been caused by the giving way of the nervous substance within.

On the other hand an increase in the quantity of the nervous substance, or an enlargement of the brain or spinal cord, consequent on an undue injection of their bloodvessels, is invariably accompanied with a diminution in the quantity of this fluid or with the total absence of it. In hypertrophy of the brain no fluid is found in the subarachnoid space, and very little or none in the ventricles. In cases of tumour of the brain encroaching upon the cranial cavity, we find no fluid; and the same is observed where chronic inflammation of the brain has given rise to a new deposit which increases the bulk and the density of the cranial contents. In all cases where a considerable quantity of fluid has accumulated *within* the ventricles, that upon the surface is either greatly diminished or entirely disappears. In the ordinary hydrocephalus internus of children fluid is never found on the exterior of the brain.

When an arrest in the developement of any portion of the cerebro-spinal axis has taken place, the space which ought to be occupied by the organ of imperfect growth is filled by liquid. In examining the heads of idiots we always find a considerable quantity of subarachnoid fluid, either general, or partial if a portion only of the brain be deficient. Or if any portion of the wall of the cranio-spinal cavity be defective, the contained viscus is protected by the accumulation of an increased quantity of liquid in the situation of the deficiency. Hence the explanation of those watery tumours which occur over various regions of the spine, in cases of *spina bifida,* in which the accumulation of water is favoured by the absence of the resisting osseous wall of the spine for a greater or less extent. And similar tumours are found projecting from the cranium, being occasioned by a protrusion of the cranial meninges through a congenital aperture, containing fluid and sometimes a portion of the encephalon itself.

Enough has been said to show, that the preternatural increase of this fluid should in general be regarded as secondary to and consequent upon the diminished size of the cerebro-spinal centre itself, and that it has most probably little or nothing to do with the manifestation of peculiar symptoms during life in the great majority of instances. Whatever be the immediate cause of the shrinking of the cerebro-spinal centre or of any portion of it, the increase of the fluid goes on *pari passu*, and in a quantity duly proportionate to the decrease of its bulk, so that it is in the highest degree improbable that, in such cases as I have enumerated, the nervous centre experiences any increased degree of pressure beyond that which it bears in the normal state. If, however, the fluid, either within or without the brain, were to increase, while that organ itself either preserved the same bulk or became enlarged, it is plain that it must experience an increased degree of compression, which doubtless would produce serious symptoms. This very rarely happens, according to my experience, as regards the sub-arachnoid fluid on the exterior of the brain : we more frequently meet with an increase of the fluid within the ventricles, and, in such cases, we shall find evidence of the compression in a manifestly greater firmness and density of its structure, and in this fact, that the lateral ventricles, when laid open by a horizontal section, do not collapse, as in the ordinary state of the brain, but remain quite patulous, owing to the firmness and density of their walls. And this patulous state of the ventricles may be regarded as a good indication that the fluid, collected in them, had for some time occasioned a preternatural amount of pressure.*

* See an important paper by the late Dr. Sims on serous effusion in the brain, Med. Chir. Trans. vol. xix.

Majendie infers, and as it appears to me with justice, that the cerebro-spinal fluid is secreted from the vessels of the pia mater. He states that, when a portion of the pia mater is exposed in a living animal, " an attentive eye may observe the transpiration of a liquid which evaporates, it is true, almost as soon as it appears, but which is sufficient to prevent the drying of the membrane." " To render this phenomenon of vital physics still more manifest," he adds, " it is necessary to inject a certain quantity of water, at 30° R., into the veins of the animal which is subjected to the experiment; immediately the liquid exhalation of the pia mater takes place in a more rapid manner, and consequently becomes more apparent." We ought to be content with M. Majendie's statement respecting this experiment: the point in question is by no means of sufficient consequence to warrant the repetition of so cruel an experiment.

Majendie's experiments have demonstrated further that this fluid can be as quickly regenerated as the aqueous humour of the eye. He found that on puncturing the theca of the spinal cord, and perforating both layers of arachnoid membrane, the fluid quickly escapes at first as a fine continuous jet, and afterwards *per saltum* in correspondence with the efforts of expiration. If the orifice be closed up and the animal left to go at large for twenty-four hours, the fluid is reproduced in as considerable quantity as before the first experiment.

What has been described as the movement of this liquid consists in an alternate elevation and collapse synchronous with expiration and inspiration, seen only when a portion of the cranio-spinal wall has been removed, and caused by the repletion of the venous system of the spine which occurs in the former state of the respiratory movements, and its collapse which takes place in the latter. The distended spinal veins compress the cerebro-spinal fluid, and cause it to rise towards the head in expiration; their collapse in inspiration favours the movement of the fluid in the contrary direction. We have no evidence from experiment or direct observation that there is any movement in the fluid of the ventricles; but the discovery of cilia upon the inner surface of these cavities seems to indicate that this fluid is not quite stationary within them.

The following account of the physical and chemical properties of the cerebro-spinal fluid is derived from Majendie's researches. When removed from the body a few moments after death, this fluid is remarkably limpid, and may be compared in this respect to the aqueous humour of the eye; sometimes it has a slightly yellowish tinge. In temperature it ranks among the hottest parts of the body. It has a sickly odour and a saltish taste; it is alkaline, restoring the blue colour of reddened litmus. Lassaigne's analysis of the human fluid yielded the following result.

Water	98.564
Albumen	0.088
Osmazome	0.474
Hydrochlorate of soda and of potass	0.801
Animal matter and phosphate of soda	0.036
Carbonate of soda and phosphate of lime	0.017
	99.980

According to M. Couerbe, some of the secondary organic products which he has obtained from the brain are to be found in this fluid. The following constituents are enumerated by this chemist: 1. an animal matter insoluble in alcohol and ether, but soluble in alkalis; 2. albumen; 3. cholesterine; 4. cerebrote; 5. chloride of sodium; 6. phosphate of lime; 7. salts of potass; 8. salts of magnesia.

What is the use of the cerebro-spinal fluid? An obvious mechanical use of this fluid is to protect the nervous centres with which it lies in immediate contact. By the interposition of a liquid medium between the nervous mass and the wall of the cavity in which it is placed, provision is made against a too ready conduction of vibrations from the one to the other. Were these centres surrounded by material of one kind only, the slightest vibrations or shocks would be continually felt, but when different materials on different planes are used, the surest means are provided to favour the dispersion of such vibrations.

The nervous mass floats in the midst of this fluid, being maintained in equilibrio in it by its uniform pressure on all sides, and the spinal cord, as we shall find by-and-bye, is supported by an additional mechanism which prevents its lateral displacement.

By its accumulation at the base of the brain, this fluid must protect the larger vessels and the nerves situate there from the unequal pressure of neighbouring parts.

It is not improbable also that this fluid may contribute to the nutrition of the brain and spinal cord, by holding in solution their proper nutrient elements preparatory to their absorption or addition to the nervous masses themselves; and this view would receive great support if Couerbe's analysis, which detects some of these elementary matters in the fluid, should be confirmed by the observations of other chemists. Nor must we omit to notice here, the fact ascertained by Majendie, that when certain substances which find their way readily into the blood have been injected into the veins, they may be soon after detected in this fluid, such as iodide of potassium.

Majendie observed serious symptoms to ensue upon the removal of this fluid from living dogs, but it is impossible to ascribe such symptoms solely to this cause, for the introduction of air into the sub-arachnoid cavity, the disturbance and consequent irritation to which the nervous centres must necessarily be exposed in the performance of the experiment, ought fairly to be considered to have a share, and that not an inconsiderable one, in any impairment of the nervous function that might become apparent. The sudden removal of the fluid brings on fainting or even death, effects due to shock, and analogous to those which result from the sudden removal of dropsical fluid in particular cavities, when the organs and the circulation in them have become adapted to its pressure, as in cases of ascites, hydrothorax, &c.

The interior of the arachnoid sac is moistened by an exhalation of a similar kind to that which is found in the other serous membranes. Accumulations of fluid in the arachnoid sac, however, are of very rare occurrence.

Of the glandulæ Pacchioni.—To these bodies we have already had occasion to refer in the description of the sinuses. We proceed now with a more special notice of them.

These bodies were first formally described by Pacchioni, and were regarded by him as conglobate glands of the dura mater, from which lymphatics proceeded to the pia mater.* They have been recognized by all subsequent anatomists under the name here assigned to them, although the idea of their physiological office suggested by Pacchioni has not met with general acceptation. Bichat suggested a more appropriate and scientific appellation in that of *cerebral granulations*. No anatomists have investigated the history of these bodies so extensively as the brothers Wenzel.†

* Ant. Pacchioni diss. epistolaris ad Luc. Schroeckhium de glandulis conglobatis duræ meningis humanæ, &c. &c. Rom. 1705, et Opusculum Anatomicum de durâ meninge, in Opera Omnia. Rom. 1741.

† Wenzel, de penitiori cerebri structura. Tubingæ, 1812.

The Pacchionian bodies are found principally along the edge of the great hemispheres of the brain on either side of the great longitudinal fissure. Here, in general, they cause the obliteration of the sac of the arachnoid for a greater or less distance by producing adhesion between the visceral layer of that membrane and that portion of its parietal layer which adheres to the angle along the superior border of the falx cerebri. In cases where these bodies are numerous and well developed, it is found very difficult to separate the dura mater from the subjacent arachnoid by reason of the firmness of the adhesion effected by them; and when this adhesion exists, the corresponding surface of the dura mater has generally a very complicated cribriform appearance. The extent of surface which they occupy is very variable. Sometimes, but rarely, they extend along the entire edge of each cerebral hemisphere; generally they occupy its central part for an extent of from one to three inches. Very frequently they extend outwards over the upper surface of the cerebral hemispheres, rarely beyond half an inch or an inch. The arachnoid membrane in their immediate vicinity is always opaque.

Bodies, somewhat similar, are also found occasionally on the choroid plexuses of the lateral ventricles. Very frequently we meet with granulations of a like kind in the fringe-like process of pia mater which descends from the velum interpositum to surround the pineal gland, and also upon the little processes of that membrane which go under the name of choroid plexuses of the fourth ventricle.

Wherever these bodies are found, they show a remarkable tendency to congregate in clusters around venous trunks. In examining them along the edges of the hemispheres, we find that they are most numerous around the veins which pass from the pia mater in that situation into the superior longitudinal sinus. This tendency, probably, explains the occurrence of these bodies in some of the sinuses. They are most commonly met with in the superior longitudinal sinus, as already stated; they are also found in the lateral sinuses, and sometimes but rarely in the straight sinus. In all these situations these bodies appear to stand in a similar relation to the sinuses; they have penetrated the fibrous tunic of their walls, and pushed before them the inner or venous tunic.

In point of size and shape the Pacchionian bodies resemble minute granulations; their colour is white, like that of coagulable lymph, and not unlike that which is occasionally seen upon serous surfaces after chronic inflammation. A granular lymph, taking somewhat a similar form, is occasionally seen on the mucous membrane of the rectum after dysentery. At some parts the granulations appear simply as isolated elevations of the surface of the arachnoid membrane. At others they are collected in clusters round a common stem; and when the membrane is removed and floated in water, this bothryoidal disposition may be well displayed. A large proportion of them cause, by their pressure, an adhesion between the opposed surfaces of arachnoid membrane; and those which are attached to a stem are the most likely to project into the interior of the sinuses.

When examined by a microscope, each of these bodies appears to consist of a mass of minute granules enclosed in a membranous sac; when the body is pediculated, its stalk exhibits a series of striæ which take the direction of its length, and probably result from longitudinal folds of the membrane which forms it. Dilute acetic acid causes them to swell and gelatinifies the bodies, and sometimes displays epithelial scales upon the surface of the membrane which covers them.

The following explanation of this structure may be offered. The primary deposit of granular lymph takes place among the vessels of the pia mater. The small bodies thus formed push the arachnoid membrane before them as a sac or covering; in some instances the granular mass is only partially covered, and then it causes merely a slight projection on the surface of the visceral layer of arachnoid; but in others the mass is completely covered, and a stalk is gradually formed; and when several granular masses have been deposited immediately contiguous to each other, they may all be attached in a cluster to the same stem. The fact that the membranous sacs of some of the bodies have epithelial particles upon their surface is sufficient proof that they are derived from arachnoid membrane. If this be admitted, then it seems impossible to come to any other conclusion than that the pia mater is the seat of the primary deposit, and this opinion is confirmed by the fact that we meet with the Pacchionian bodies on the internal processes of the pia mater, when we have no evidence of the existence of arachnoid membrane.

Or it might be conjectured that these bodies indicate a degenerate condition of the elementary particles of the superficial layer of the grey matter of certain convolutions, produced by frequent irritation.

Are the Pacchionian bodies natural or morbid structures? The great frequency with which these bodies are met with in the various situations above-mentioned, has induced many, even in the present day, to regard them as normal structures, the physiological office of which is as yet unknown. But there are many facts which strongly militate against such a conclusion. In the first place it may be observed that Pacchionian bodies never occur in the earliest periods of life. In the course of a long experience in anatomical investigations I have never seen them at a period antecedent to six years. The brothers Wenzel, who made a series of special examinations with a view to determine this question, make the following statement. In children, from birth to the third year, these bodies, if they ever occur, must be very few. From the seventh to the twentieth year they sometimes are numerous. From the latter period to the fortieth year their number is considerable, and the nearer we approach the fortieth year the greater does it become. Lastly, from the fortieth to the one hundredth year these bodies are found in great numbers.

It must be further remarked that even at those periods of life when the Pacchionian bodies are found in greatest numbers, cases frequently occur in which no trace of them can be found. There is likewise the greatest variety as to their number and size, in different individuals of the same age.

It has always occurred to me to find them most numerous in cases where I had reason to know that the brain had been subject to frequent excitement during life. In persons addicted to the excessive use of spirituous liquors, in those of irritable temperament and who had frequently been a prey to violent and exciting passions, they are almost uniformly highly developed.

The Pacchionian bodies are peculiar to the human subject. Nothing similar to them has been found in any of the inferior classes of animals.

In reference, then, to the question, what is the nature of these bodies, I have no difficulty in stating my opinion that the evidence greatly preponderates in favour of their morbid origin; that they are the product of a chronic very gradual irritation due to more or less frequent functional excitement of the brain itself. It is not unlikely that the friction to which the opposed surfaces of the arachnoid are continually subjected in the movements of the brain, especially when they are of a more rapid and violent kind, as under states of cerebral excitement, may contribute to the developement of many of the appearances connected with these bodies. The opaque spots which are of such frequent occurrence upon the surface of the heart may be quoted as an example of a morbid change, very commonly met with, and resulting probably from the friction against each other of opposed serous surfaces. Were the Pacchionian bodies normal structures, they would not be so frequently absent from brains which afforded every other indication of being in a healthy state; nor should we find opacity of the arachnoid (a decidedly unhealthy condition) so commonly coexistent with the full developement of them. Again, were they a necessary part of the healthy organism, we might expect to find them more constant as regards size, number, and the extent of surface over which they were placed.

Of the ligamentum dentatum (serrated membrane of Gordon). This structure forms a part of the mechanical arrangements connected with the spinal cord and the roots of its nerves. It is found in the subarachnoid space, adhering on the one hand to the pia mater, and, on the other, attached at certain intervals to the dura mater.

The ligamentum dentatum consists of a narrow longitudinal band, adhering by its inner straight border to the pia mater on each lateral surface of the spinal cord, midway between the anterior and posterior roots of the spinal nerves, reaching from the highest point in the cervical region down to the filiform prolongation with which it becomes incorporated. Its outer border exhibits a series of tooth-like triangular processes which are inserted by their apices into the dura mater. The first pointed process, which is longer than the rest and less triangular in shape, is inserted into the dura mater on the margin of the occipital bone, where it stands in relation with some parts of interest. The posterior root of the sub-occipital nerve, and the filaments of origin and the resultant trunk of the spinal accessory, are on a plane posterior to it. The vertebral artery and the ninth pair of nerves are anterior to it. The number of teeth varies from eighteen to twenty-two. The last is attached

THE LIGAMENTUM DENTATUM. 61

Fig. 11.

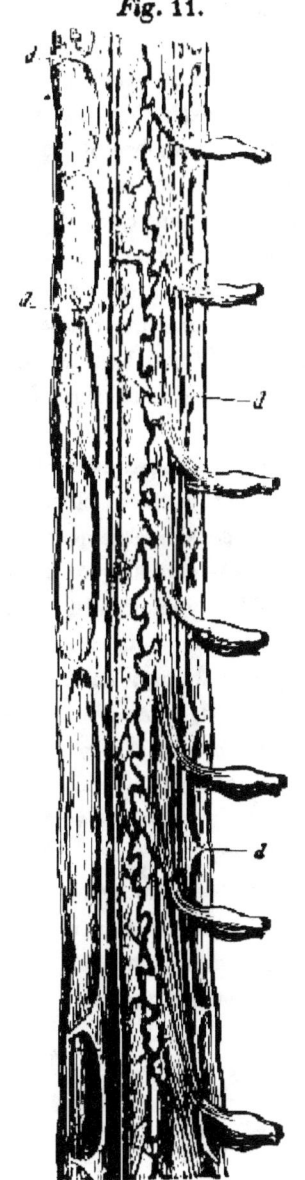

Dura mater of part of the spinal cord laid open to show the ligamentum dentatum. d d d d, dentated processes. On the right the roots of the nerves and the ganglia of the posterior roots are retained.

to the dura mater about the level of the first or second lumbar vertebra. The points of attachment are between the orifices of exit of the spinal nerves, being almost always nearer the lower than the upper nerve. The intervals between each pair of dentated processes vary in different regions of the spine as the distances between the roots of the nerves vary. At its insertion into the dura mater each process pins down the visceral and parietal layers of the arachnoid membrane, probably piercing them to reach the fibrous membrane. At its lowest part, a little above the extremity of the cord, the denticulate margin ceases, and the longitudinal portion may be traced downwards, gradually diminishing in size, along each side of the filiform prolongation of the pia mater.

The dentated ligament has to the naked eye all the characters of white fibrous tissue, of which it is chiefly composed. In its dentated processes, however, a considerable quantity of yellow fibrous tissue may be found. The similarity of its constitution with that of the pia mater evidently justifies its being regarded as a process of that membrane, and not, as some anatomists suppose, of the dura mater, with which it has a much less intimate and extensive connexion. Its anterior and posterior surfaces are uncovered by any membrane; they are smooth, and have the glistening silvery appearance of white fibrous membrane. It is evident that during life these surfaces must be bathed by the subarachnoid fluid.

The office of this remarkable structure seems evidently to be mechanical; to preserve the spinal cord in a state of equilibrium; and to prevent lateral movement of it, whilst at the same time it forms a partition between the roots of the nerves.

CHAPTER II.

General remarks on the structure of the nervous centres— Of the grey or vesicular nervous matter—Of the structure of ganglions.

THE nerves properly so called are composed exclusively of one kind of nervous substance,—namely, the fibrous nervous matter, which is disposed in bundles of peculiar fibres. It is only in the nervous centres or in continuations of them that we find an union of the white and the grey or vesicular nervous matter; and, indeed, it may be stated in general, that the peculiar and distinctive anatomical character of a nervous centre consists in this combination of the two kinds of nervous matter.

In the nervous centres the white matter exhibits, for the most part, the same essential characters of structure as in the nerves; that is to say, it is disposed in tubes containing a certain pulpy matter in them. It has been found, however, that these tubes are much more prone to become varicose under the influence of pressure or of any other disturbing cause. They are not, as in the nerves, bound together by areolar tissue, but are disposed in bundles and on different planes, with their nutrient bloodvessels ramifying among them, and in some situations the elements of the grey matter are interposed between them. Certain parts of the nervous centres are composed exclusively of white matter, as a portion of the hemispheres of the brain, and of the cerebellum, and the superficial parts of the spinal cord.

The white fibres which are found in the nervous centres may be distinguished according to their physiological office into four different kinds. Two of these are continuations of the fibres of the nerves, and serve to connect the nervous centres with other organs or textures, either by conveying the influence of the centres to them, or by propagating impressions from them to the centres. The former are called *efferent*, the latter *afferent* fibres. In addition to these, we find a third and large series of fibres, which serve to establish a connection between different centres, or between different portions of the same centre. These are called *commissural* fibres; they form a large portion of the mass of the brain and spinal cord. And Henle suggests that the brain contains a fourth series of fibres, associated with the operations of thought.

We remark in the nervous centres, especially in the brain and spinal cord, a greater difference as regards size between the different nerve tubes, than may be observed elsewhere, and it seems to be a constant character that they diminish in size as they approach and enter the grey matter.

Of the vesicular or grey nervous matter.—This form of nervous matter differs very materially in its anatomical characters from the white or fibrous. Its elements are vesicles or cells, with nuclei and nucleoli. Although this vesicular or cell form is universally prevalent, the cells present much diversity of shape, size, and colour in different centres or even in the same centre, which apparently have reference to some peculiarity of function. The most prevalent form is that of a globular vesicle, composed of a very delicate transparent membrane. Within this membrane is contained a soft minutely granular substance, which forms the principal mass of the body, *parenchymmasse* (Valentin). The grey colour of the vesicle, which becomes very manifest when a number of them is congregated together, is dependent on this granular matter. (See *fig.* 12, *a, b, c*.) When the vesicle bursts and its substance is broken up, the granular matter is diffused, and confuses and darkens the specimen under examination. Sometimes the outer vesicle is removed, the contained granular matter retaining the globular form. Within the external vesicle (*a, fig.* 12) there is ano-

Fig. 12.*

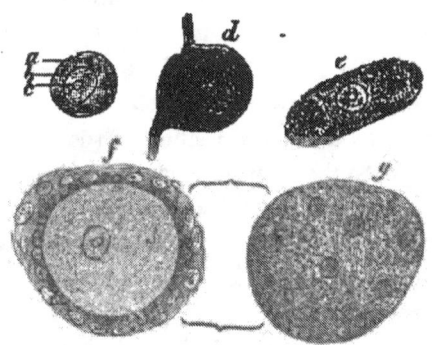

Nerve vesicles from the Gasserian ganglion of the human subject.

a, a globular vesicle with defined border; *b*, its nucleus; *c*, its nucleolus; *d*, caudate vesicle; *e*, elongated vesicle with two groups of pigment particles; *f*, vesicle surrounded by its sheath or capsule of nucleated particles; *g*, the same, the sheath only being in focus.

ther much smaller and adherent to a part of its wall, so as to be quite out of the centre of the containing vesicle. This is the *nucleus* (*b, fig.* 12). Its structure is apparently of the same nature as that of the external vesicle. The nucleus contains in its centre another minute and remarkably clear and brilliant body, also vesicular in structure. This is the *nucleolus* (*c, fig.* 12). Sometimes it is replaced by two or three much smaller but similar bodies. The softness of the vesicle admits of its yielding, whether from the disturbance occasioned in the necessary manipulation or from the pressure of the neighbouring elementary parts as it lies in its proper

* I am indebted to the accurate pencil of my friend Mr. Bowman for this illustration.

situation. Hence it is that these vesicles exhibit a considerable diversity of form.

Very frequently we observe that, besides the granular substance above described, there are certain pigment particles of large size and dark colour, which are collected into one or two roundish or oval groups, situate at or towards one or both sides of the vesicle (*fig.* 12, *e*). These masses of colouring matter sometimes occupy considerable space, and enable the observer readily to detect the position of such vesicles as contain them. When the mass of pigment is placed at one side, we may compare the containing vesicle, as Volckmann has done, to a fruit which is coloured only on that side which is exposed to the sun. The aggregation of many such vesicles at any one spot gives the nervous matter there a peculiarly dark colour. A remarkable example of this is found in that portion of the crus cerebri which is known by the name of *locus niger*.

A very interesting form of nerve-vesicle is that which exhibits the greatest departure from the globular shape by the prolongation of the wall of the outer cell into one or more tail-like processes. These bodies may, from this peculiar character, be designated *caudate nerve-vesicles*. They possess the nucleus and nucleolus, as in the more simple form, and contain one or more of the masses of colouring matter; indeed, in them the quantity of pigment is generally much more considerable than in any other form. I have noted an observation which shewed two nuclei in one vesicle. They vary much in size and shape, and so also do the processes. The largest nerve-vesicles are found among those of this description. The variety in shape may depend in some degree upon the situations from which the caudate processes take their rise. In some (*fig.* 12, *d*) they proceed from opposite poles of the vesicles; in others they arise near each other from the same region of the vesicle, and when numerous, give to it somewhat the form of a cuttle-fish with extended tentacles. In examining the structure of one of these processes, we find it evidently exactly similar to that of the matter contained in the outer vesicle, exhibiting the same minutely granular appearance. The processes are implanted in the surrounding substance, and firmly connected with it, so as to be with great difficulty separated from it. They exhibit much strength of cohesion, but are frequently broken off quite close to their points of origin, and the broken ends present a distinctly lacerated edge (*d, fig.* 12). More rarely we are able to trace these processes to a considerable distance, and then we observe them to bifurcate or even to subdivide further, and to terminate in exceedingly fine transparent fibres, the connexion of which with the other elements of the nervous matter has not yet been ascertained.*

It is in vain, in the present state of our knowledge, to speculate upon the use of these caudate processes. Do they constitute a bond of union between the nerve-vesicles and certain nerve-tubes? or are they commissural fibres serving to connect the grey substance of different portions of the nervous centres? Until a more extended research has made us better acquainted with the peculiarities of these vesicles in various localities, it would be premature to offer any conjecture concerning their precise relation to the other elements of the nervous centres. They exist, with different degrees of developement, in the locus niger of the crus cerebri, in the laminæ of the cerebellum, in the grey matter of the spinal cord and medulla oblongata, and in the ganglions, and in the grey substance of the cerebral convolutions, in which latter situation they are generally of small size.

When a portion of grey matter from a convolution of the

* See a beautiful illustration of one of the largest of these vesicles in the second part of Mr. Bowman's and my work on Physiological Anatomy and Physiology. (*Fig.* 56, p. 214.).

brain is examined with a high power in the microscope, we observe it to consist chiefly of a mass of granular matter, in which nerve-vesicles are imbedded with considerable intervals between them. Henle states, with much truth, that the superficial part of the grey matter of the convolutions seems almost entirely composed of finely granular substance, in which lie, scattered here and there, several clear vesicles which, as he remarks, look almost like openings or circular solutions of continuity (*fig.* 13). In the middle portion the vesicles appear larger, and the granular matter becomes less abundant, and on the most deep-seated plane the nerve-vesicles are much increased in size and lie in closer juxtaposition, being, however, covered by a thin layer of granular matter, which forms a sheath to each vesicle. Nerve-tubes are found throughout the whole depth of the grey matter. Those in the most superficial layer are extremely fine and varicose, and seem to correspond in number and situation to the vesicles. For

Grey substance from the surface of the cerebral hemisphere of a full-grown rabbit treated with dilute acetic acid. (After Henle.)

a, nerve-vesicle; b, a similar one with two nuclei; c, another viewed along its edge; d, vesicles indistinctly apparent; e, granular matter.

wherever there is a nerve-vesicle, we find an extremely fine varicose nerve-tube apparently adherent to it.

In the grey matter of the ganglions we find that the vesicles are also deposited in granular matter, which surrounds each of them as a sheath (*fig.* 12, *f*, *g*), completely investing it on every side, and separating it from the neighbouring ones. Here, however, the sheath is formed not only of a finely granular matter, but also of numerous bodies which resemble nuclei or cytoblasts, and this sheath invests both the globular variety of nerve-vesicles and the caudate ones. Nerve-tubes lie in immediate connexion with these vesicles, and sometimes entwine themselves around them, and seem to indent their sheaths (*fig.* 16).

Other vesicles, much more simple in form, are found in the grey matter in certain situations. The outer layer of the optic thalamus, according to Henle, contains only small homogeneous globules, analogous to the nuclei of the ganglionic globules, in immediate apposition with each other, and towards which the tubes seem to ascend in the vertical direction. Purkinje states, that a similar layer is met with in the cortical substance of the brain quite close to the medullary substance.[*]
I find a layer of similar particles in the grey matter of the cerebellic laminæ. And, according to the report of Valentin,[†] Purkinje has found the interior of the ventricles in the normal state covered by an oily matter, which consists of distinct, large, transparent globules, free and lying near each other. A similar layer has been found by him in the interior of the fifth ventricle. The cavity of the rhomboidal sinus in Birds likewise contains a gelatinous mass, which consists of large globules lying close to each other.

Developement of grey matter.—In the perfect nerve-vesicle,

---

[*] Henle, loc. cit.   [†] Uber den Verlauf, &c.

the cell form of primitive developement is persistent. We have the nucleolus and nucleus (cytoblast) and the cell; and, according to Schwann, the only change which the full-grown cell exhibits consists in an increase of size and in the developement of the pigmentary granules within. The following is Valentin's description of the developement of a nerve-vesicle. In the very young embryos of Mammalia, as the sheep or calf, the cerebral mass in the course of formation contains, in the midst of a liquid and transparent blastema, transparent cells, with a reddish yellow nucleus. The wall of the cells is very thin and simple; their contents are colourless, transparent, homogeneous, and manifestly liquid; the nucleus, with well-defined contour, is generally round, sometimes central, at other times excentric, solid, and nearly of the same colour as the corpuscles of the blood. Around these primitive cells of the central nervous system, which we find likewise formed after the same type in the spinal cord, a finely granular mass becomes deposited, which probably is not at first surrounded by an enveloping cell-membrane. At this early period of formation the primitive cell still preserves its first delicacy to such a degree that the action of water causes it to burst immediately. This rupture of its membrane and effusion of its contents often take place so suddenly and quickly that they can be perceived only by the movement of the nucleus, which is the consequence of it. * * * In proportion as the granular mass contracts itself within certain limits, (sich immer mehr abgrenzt,) a cell-membrane probably becomes developed around it, so that the vesicle gradually acquires its precise form and size, and its contents their proper characters, which belong to a fully formed central nervous corpuscule.* Valentin compares the developement of these vesicles to that of the ovum. The nucleolus of the nerve-vesicle is always first formed, then

* Valentin, in Soemmering vom Baue, &c. t. iv. § 24.

around it the primitive cell, and around this the outer cell. This process resembles exactly that which takes place during the formation of the ovum, for the germ corresponds to the nucleolus, the germinal vesicle to the nucleus, the yolk to the contents of the outer enveloping cell, and the vitelline membrane to the delicate wall of this cell, supposing that this latter membrane always exists.*

The great simplicity in the form of the elements of the grey nervous matter is one of its most remarkable characteristics. That a tissue, which plays the prominent part in the nervous actions, whether they are prompted by mental change, or are purely corporeal, should exhibit scarcely any more complexity of structure than that which is found in the simplest animal or vegetable textures, or in structures that have not passed their earliest phase of developement, is an anatomical fact pregnant with great physiological interest. Have this simplicity of form and delicacy of structure reference to the celerity of the nervous actions? or to that proneness to change which must be induced by the constant and unceasing round of impressions which the grey matter must receive from the ordinary nutrient actions that are going on in the body, as well as from the continual action of thought? If, according to common acceptation, we admit that the mind has some immediate connexion with the cerebral convolutions, it may well be imagined that no part of the frame can be the seat of such active change, from its being on the one hand the recipient of impressions from the body, and, on the other, from an association with the psychical principle so intimate that probably, under ordinary circumstances, an affection of the one cannot occur without being communicated to and producing a change in the other.

Another curious fact, in connexion with the intimate struc-

* Loc. cit. § 25.

ture of the grey nervous matter, is the large quantity of pigment or colouring matter which exists in it, and which appears to form one of its essential constituents, more abundant in some situations than in others, but present in all. We are utterly ignorant of the design of this peculiarity of structure. If this pigment bear any resemblance of chemical composition to the colouring matter of the blood, *hæmatosine*,—and it is not improbable that it does,—an increased interest attaches to the practical importance of minute attention, on the part of practitioners, to avail themselves of all the means which are capable of improving that important element of the nutrient fluid both in quantity and quality, for it is most reasonable to presume that the pigment of the nervous matter would derive its nourishment from that of the blood.

It may be further remarked that pigment occurs in connexion with the nervous system in another form besides that of incorporation with its elementary particles, that is, upon the exterior of parts of the nervous centres or of particular nerves. Examples of this may be referred to in the case of the olfactory nerve of the sheep and of other Mammalia, the bulb of which is surrounded by black pigment connected with the pia mater. It is also found sometimes on the pia mater of the spinal cord of the human subject. Valentin, who delineates a magnified view of this pigment, states that it occurs chiefly in the cervical region. In frogs, the whole spinal cord and encephalon are covered with a silvery pigment interspersed with black. The same occurs in fishes. The black pigment in connexion with the retina has an obvious use. On the choroid gland of fishes, which lies immediately contiguous to the retina and surrounds the optic nerve, there is a silvery membrane which contains a quantity of the same kind of pigment as that alluded to upon their nervous centres. On some of the ganglia of the invertebrata particles of pigment are likewise found.

Of the structure of ganglions.—The description of the minute anatomy of ganglions as well as of all other nervous centres may be regarded as the solution of the following problem: to determine the relation which the fibrous substance of these centres bears to the vesicular matter on the one hand, and to the nervous trunks connected with them on the other hand.

The fibrous substance of the ganglions consists of a series of minute nerve-tubes, as well as of some gelatinous fibres, which are continuous with those which exist in the nerves themselves. If we trace a nerve into a ganglion, it is found to break up into its component nerve-tubes, and it does so by a separation of the tubules within into smaller bundles, or single tubes. Sometimes adjoining bundles interlace, each yielding to its neighbour one or more tubes. The nerves which

Fig. 14.

Second abdominal ganglion of a greenfinch, slightly compressed under the compressor. The course of the nerve-tubes only is represented.

a, fibres passing in; *b*, emerging fibres; *c*, surrounding fibres. The meshes for the reception of ganglion globules are shown.

emerge from the ganglia derive their component nerve-tubes from different bundles, so that the same kind of interchange of tubules, which we have noticed as taking place in plexuses, occurs also in ganglia. The emerging nerves result from a further subdivision and greater intermixture of the bundles of nerve-tubes which enter the ganglions. The arrangement is well shown in *fig.* 14, where the nerve (*a*), which enters the ganglion, may be seen breaking up into a plexus, from which three branches (*b*, *b*, *b*) emerge, and it may be observed that these emerging nerves derive nerve-tubes from very different and opposite parts of the ganglionic plexus. In the meshes, which are left between the interlacing nerve-tubes, the ganglionic globules or nerve-vesicles are situate (*figs.* 14, 15). Certain fibres, according to Valentin, travel round the margin of the ganglion, and to these he gives the name of *umspinnende Fasern*, surrounding fibres, and some fibres pass from them to the more central ones, or from the latter to the former. Nerve-vesicles exist at the circumference of the

Fig. 15.

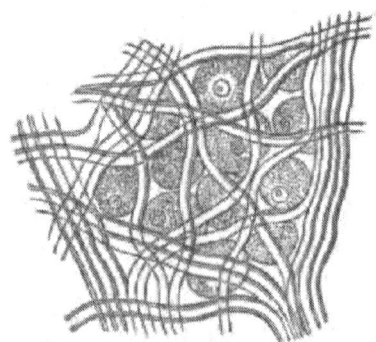

A small piece of the otic ganglion of the sheep, slightly compressed, showing the interlacement of the internal fibres and the grey matter. (After Valentin.)

ganglion as well as in its interior, and to them is due the peculiar grey colour of that body.

The best mode of examining these points is to select the smallest ganglia of very small animals, birds, mice, &c.; these, when subjected to compression, become very transparent, and display much of their intrinsic arrangement. Or thin slices of large ganglia may be placed under the microscope, and when torn up by needles the disposition of the nerve-vesicles and the caudate processes, when present, are rendered visible. None is more suitable for this purpose than the Casserian ganglion of the fifth nerve, which from the absence of a dense sheath and its greater looseness of texture is more easily examined.

It is a highly important problem, in minute anatomy, to determine whether there are any nerve-tubes which terminate in the grey matter of the ganglion, or originate in it,—which in short are not continued through the ganglion. At present we are unable to state further than that the tubes appear to have an intimate connection with the nerve-vesicles wherever the latter may be found, and that they often appear to be continuous with the sheaths of the nerve-vesicles.

Fig. 16.

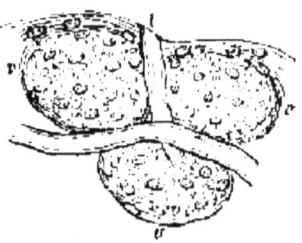

Nerve vesicles from the Gasserian ganglion covered by their sheaths of nucleated particles, to shew the intimate relation of the nerve-tubes to them.

t, t, nerve-tubes; *v, v,* vesicles.

There does not appear to be any material difference of structure between the ganglions of the sympathetic and those of the cerebro-spinal system, excepting, as Henle states, the existence of a greater number of gelatinous fibres in the former.

CHAPTER III.

Of the cerebro-spinal centre in general—Of the spinal cord— Origin of spinal nerves—Characters of the spinal nerves in each region.

THE nervous mass which occupies the cavities of the cranium and spine doubtless constitutes one great centre, as there is a perfect continuity throughout all its parts. But the differences of external form and characters in some regions of it, and the obvious diversity of endowment of the nerves connected with certain portions, denote and justify an anatomical as well as a physiological subdivision of it into segments, each of which is a centre of nervous action independent of the rest, yet so connected with them that the functions of all are made to harmonize in the most perfect manner.

The subdivision which the external anatomy indicates, although not perfectly coincident with that which the differences of function would suggest, has been so long sanctioned by usage and is so convenient for description, that no advantage would be gained by adopting any other. Our description of the cerebro-spinal centre, or axis as it has also been called, will be given under the following heads: 1. the spinal cord; 2. the encephalon, including *a*, the medulla oblongata; *b*, the mesocephale; *c*, the cerebellum; *d*, the cerebrum.

OF THE SPINAL CORD.—Syn. *Spinal marrow, Medulla spinalis;* Fr. *La moëlle epinière;* Germ. *Das Rückenmark.* The following are the anatomical limits which may be assigned to the spinal cord. It occupies a large portion of the spinal canal, terminating inferiorly at a point which, in different subjects, ranges between the last dorsal and the second lumbar vertebra. Below this point the sheath formed by the dura mater contains that leash of nerves which is called the *cauda equina,* in the centre of which lies the filiform prolongation or process of the pia mater. The superior limit of the spinal cord is marked by the plane which lies between the occipital foramen and the first vertebra of the neck. A section made in the direction of this plane separates the spinal cord from the medulla oblongata. Immediately above this plane the decussation of fibres of the anterior pyramids takes place, and may be regarded as the natural inferior limit to the medulla oblongata.

Such is the position of the spinal cord in the adult. In the fœtus at the third month of intra-uterine life, it occupies the whole spinal canal, and extends quite to the point of the sacrum. At this early period the os coccygis consists of seven vertebræ. Coincident with the reduction to its normal number of segments, is the retraction of the spinal cord within the spinal canal. If the ascent of the cord be arrested, the fœtus is born with a tail, for the changes of the coccyx become arrested also. It is remarkable that among the inferior animals there is a direct proportion between the length of the spinal cord and that of the tail. The shorter the former, or the higher in the spinal canal it may be, the less will be the latter. In animals with long tails there is no *cauda equina,* as is the case in the ox, the horse, the squirrel, &c. and the opposite is likewise true, namely, that in animals with a short tail the spinal cord is much shorter and is placed higher up in the spinal canal. In the embryo of the bat, which has a tail, the spinal cord extends downwards, but when it loses its tail the cord appears to occupy a much smaller portion of the spinal canal.

In the tadpole of the frog, likewise, the spinal cord extends into the tail, but when the tail has disappeared the cord occupies only a portion of the spinal canal.*

In point of shape, the spinal cord is cylindroid, slightly flattened on its anterior and posterior surfaces, more so on the former than on the latter. At its inferior extremity it gradually tapers to a point. Sometimes, however, we observe a small tubercle immediately above this pointed extremity, situated on the posterior surface. The perfect cylindrical form of the cord is destroyed, not only by this pointed termination and the flattening before and behind, but likewise by a marked change of dimensions in certain regions. In the cervical region we observe a distinct swelling or enlargement, which begins a short distance beneath the medulla oblongata, and gradually passes into the dorsal portion, which is the smallest, as well as the most cylindrical part of the cord. This cervical enlargement *(intumescentia cervicalis)* begins opposite the third cervical vertebra, and ends about the third dorsal. The cord continues of a cylindrical form as low as about the ninth or tenth dorsal vertebra, and then passes into the lumbar swelling *(intumescentia lumbalis vel cruralis)*, which occupies a space corresponding to about two vertebræ. This swelling is both shorter and of less diameter than that in the region of the neck. The inferior extremity of the spinal cord tapers rather suddenly, and at its point is enclosed in the commencement of the filiform prolongation of the pia mater.

The bulk of the spinal cord is in the direct ratio to that of the body throughout the vertebrate series. And not only is this true with regard to the whole cord, but with respect to its segments. For when any segment supplies nerves to a

* Cuvier's Report upon Serres' work, Sur l'Anat. Comp. du Cerveau. Par. 1824.

greater sentient surface, or to more numerous or more powerful muscles than another, it exhibits a proportionally greater size. It is thus that we may satisfactorily explain the occurrence of the cervical and lumbar enlargements. Both supply nerves to the extremities, whilst the dorsal portion furnishes them only to the trunk. The upper extremities enjoy, in part, a high degree of tactile sensibility, and they possess great power and extent of muscular movement. That portion of the cord therefore from which the nerves to the upper extremities proceed is larger in every way than that which supplies the lower extremities, which, although provided with large and powerful muscles, do not enjoy such a range or variety of motion as the upper extremities, nor are they endowed with so exquisite a sensibility.

There are many facts among the lower animals which illustrate and confirm this law. Thus, in animals which have no limbs, as serpents, the cord is of equal size throughout, excepting at its pointed extremity. It is said that in the fœtus, before the developement of the limbs, no distinction of size can be discovered in the cord, and in persons in whom an arrest in the developement of the upper extremities has taken place, there is no cervical enlargement. Cruveilhier refers to the case of the tortoise as strongly confirmatory of this law. That portion of the spinal cord which corresponds to the carapace, which is equally devoid of sensation and motion, is reduced to a mere thread, whilst those segments between which it lies, and from which the nerves of the extremities emanate, are of size duly proportionate to their muscular activity and their sensibility. In Fishes, the enlargements correspond to the fins which are possessed of greatest muscular power. In the gurnard there exist certain very remarkable ganglionic swellings, situate on the posterior part of the cervical segment of the cord. With these swellings nerves are connected, which are distributed to organs placed immediately behind the head on the lower part of the body. These organs are endowed with much tactile sensibility, and seem to serve the office of feelers, as the animal gropes along the bottom of the sea.

The length of the spinal cord in the adult is from sixteen to eighteen inches, according to the statement of Cruveilhier. Its circumference measures twelve lines at the smallest and eighteen lines at the most voluminous part. Chaussier states that its weight is from the nineteenth to the twenty-fifth part of that of the brain in the adult, and about the fortieth part in the new-born infant. The actual weight of the spinal cord in an adult male may be stated to be a little more than one ounce.

We may here again notice the interesting fact that there is a great disproportion between the size of the spinal cord and that of the vertebral canal, and that consequently a considerable space is found between the cord and its membranes which is occupied by the cerebro-spinal fluid.

The consistence of the medullary substance of the spinal cord, in the fresh state, is of much greater firmness than that of the brain. This lasts, however, but for a very short time, for decomposition sets in quickly, and then the cord acquires a pultaceous consistence, and the nervous matter may be easily squeezed out of the sheath of pia mater in which it is enclosed.

The pia mater adheres very closely to the surface of the cord, as intimately as the neurilemma to a nerve. In order to examine the surface of the cord, the best mode of proceeding is to dissect off the pia mater carefully, the cord having been fixed under water. The dissector will then perceive that numerous minute vessels, accompanied by delicate processes of the membrane, penetrate the cord at all points from the deep surface of the pia mater, and to this is due the adhesion of this membrane above-mentioned. This arrangement may also be shewn by dissolving out the nervous matter through the action of liquor potassæ. The prolongations from the deep surface of the sheath may be displayed by floating the preparation in water.

The spinal cord is penetrated both on its anterior and posterior aspect by fissures, each of which corresponds to the median plane. They are separated from each other by a transverse bilaminate partition of white and grey matter, of which the grey layer is posterior. This serves to connect the equal and symmetrical portions into which the cord is divided by these fissures.

The *anterior* fissure is very distinct and easily demonstrated. A folded portion of the pia mater may be traced into it down to the commissure. The edges of this fold, as it enters the fissure, are connected by a band of white fibrous tissue, which may be traced through the whole length of the cord on the exterior of the pia mater, and indicates precisely the position of the anterior fissure, and which covers the anterior spinal artery. When this fold is carefully removed, the floor of the fissure becomes apparent, formed of a lamina of white nervous matter. This layer is perforated by a great number of minute orifices, which give to it quite the cribriform character, and are for the reception and transmission of bloodvessels. In many parts the layer appears to be composed of oblique and decussating fibres, as if the same kind of decussation which occurs at the lower part of the medulla oblongata extended through the whole length of the cord. There is not, however, any real decussation : the appearance of it results from the foramina not being always on the same level. For, in those places where they lie quite on a level with each other, no one could suppose that such an arrangement of fibres existed. Here, as elsewhere, the fibres assume the transverse direction. The depth of the anterior fissure is not the same all down the cord; it gradually diminishes towards its lowest point; its deepest part, however, corresponds to the cervical enlargement, and here it is about one-third of the thickness of the cord measured from before backwards.

The whole cribriform layer which forms the floor of the anterior fissure constitutes a commissure between the lateral halves of the cord in their whole length. It is called the *anterior* or *white commissure* of the cord.

The *posterior* fissure is very much finer and more difficult to demonstrate than the anterior. It is not penetrated by a *fold* of the pia mater: a single and very delicate layer of that membrane is continued from its deep surface down to the floor of the fissure. At this situation the spinal pia mater assumes the appearance and character of that of the brain. Here and there, within the fissure, the membrane appears interrupted and the vessels extremely few, and in such situations the fissure becomes very indistinct and difficult to recognise. The process of pia mater becomes extremely delicate towards the lowest extremity of the cord. The posterior fissure is deeper than the anterior. Through a great part of its course it is equal to fully one-half of the thickness of the cord; in the lumbar region, however, its depth is very much less. Its floor is formed by a layer of grey matter, which connects the cineritious matter of each lateral half of the cord, and which is called *the grey commissure*. In the lumbar region, however, it does not appear to reach the grey commissure.

Arnold denies the continuity of the posterior fissure through the greater part of the cervical and dorsal regions. According to his figures it ceases on a level with the second cervical nerve and reappears about the second dorsal vertebra. This does not at all accord with my observation, nor is it confirmed by any anatomist that I know of. It appears to me that the continuity of the fissure might be more properly questioned at the lowest third of the cord, where it is often so feebly developed as to elude detection. In three out of four specimens now before me, the fissure is sufficiently distinctly marked down to quite the lowest extremity of the cord, and the posterior columns separate readily from each other along it. In the fourth, which is quite recent, the fissure at the lowest part of the cord is only to be distinguished here and there by a solitary red vessel passing to the grey commissure, and is most distinct in the cervical region. Those specimens which shew it well have been hardened in a preserving liquid, which, by constringing the substance of the posterior columns, renders the fissure much more distinct. There seems little doubt that the posterior columns have no connexion with each other as far down as the lumbar region. Below that, however, it is not improbable that they may be united across the middle line, and that to this cause the indistinctness of the posterior fissure may be due. And this anatomical fact may be quoted as, in some degree, adverse to the theory which regards these columns as sensitive: for were they columns of sensation, it is probable that the preservation of their distinctness would have been more fully provided for.

The anterior and posterior fissures, as Cruveilhier remarks, leaving on each side a perfectly symmetrical organ, serve to demonstrate the existence of two spinal cords, one for each side of the body, and both presenting a perfect resemblance of form and structure.

There are no other fissures in the cord besides those just described. Several anatomists regard the lines of origin of the anterior and posterior roots of the nerves as constituting distinct fissures. But a little careful examination will readily shew that there is no real separation of the nervous substance of the cord corresponding to these lines, and that there

SUBDIVISION OF THE CORD.

is no anatomical indication of a subdivision into columns or segments in connexion with them. When the roots of the nerves have been removed on each side, nothing is seen but a series of foramina or depressions corresponding to the points of emergence of the nerve-fibres, of which the roots are composed.

The most natural subdivision of the spinal cord is that which is obviously indicated by its internal structure. In examining a transverse section (*fig.* 17), we observe that the interior of

Fig. 17.

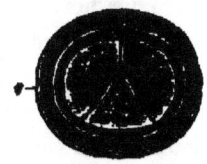

each lateral portion is occupied by grey matter, disposed somewhat in a crescentic form. The concavity of the crescent is directed outwards: its anterior extremity is thick, and is separated from the surface of the cord by a considerable layer of white nervous substance. The grey matter is prolonged backwards and outwards in the form of a narrow horn, which reaches quite to the surface of the cord, and near the surface experiences a slight enlargement. This posterior horn constitutes, on each side, a natural boundary between the two columns of which each lateral half of the cord consists. All that is situate in front of the posterior horns is called the *antero-lateral* column, and this comprehends the white or fibrous matter forming the sides and front of the semi-cord, limited anteriorly by the anterior fissure and posteriorly by the posterior roots of the nerves. The posterior column is situate behind the posterior horn of grey matter, and is separated from its fellow of the opposite side by the posterior fissure.

According to this view, then, the spinal cord will be found to consist of four columns, between which an obvious line of demarcation exists throughout the whole length of the organ. These are two *antero-lateral* columns and two *posterior* columns. The former constitute by far the largest proportion of the white substance of the cord, and envelope the anterior obtuse portion or horn of the grey matter. The white commissure at the bottom of the anterior fissure unites them. The anterior roots of the nerves are connected with them, and the posterior roots adhere to them when the cord is split up along the plane of the posterior horn. The posterior columns are small, in section triangular, placed in apposition with each other by their inner surfaces. Their apices are directed forwards, and their bases, which are slightly curvilinear, backwards. No distinct commissure of white fibres can be detected uniting these columns, save, perhaps, in the lumbar region. The connexion of the posterior roots of the nerves with them must necessarily be very slight, as they invariably separate from them in the longitudinal splitting of the cord.

The arrangement of the grey matter in the cord, as already partly explained, is as follows:—In each lateral half there is a portion of grey matter, which is crescentic in form, having its concavity directed outwards and its convexity inwards towards its fellow of the opposite side. The anterior extremity or horn of the crescent is thick and roundish, and its margin has a jagged or serrated appearance, which is more conspicuous in some situations than in others. The posterior horn is directed backwards and a little outwards: it reaches the surface of the cord, and near its posterior extremity it presents a swollen or enlarged portion, which differs in colour and consistence from the rest of the crescent, being somewhat paler and softer. This portion of grey matter has been called by Rolando *substantia cinerea gelatinosa*. It is that part of the grey matter which appears to be more immediately connected with the posterior roots of the nerves.

There is an exact symmetry between the grey crescents of opposite sides, and they are united by means of the grey commissure, a layer which extends between the two crescentic portions, being attached very nearly to the central point of each. This commissure, then, when examined in its length, forms a vertical plane of grey matter, extending throughout the whole of the cord. The lateral portions are solid masses of grey matter, with which the nerve-tubes of the white substance freely intermingle, and in which, as in the grey matter elsewhere, very numerous bloodvessels ramify. There seem to be no good grounds for the opinion advanced by Mayo that these crescentic portions are hollow capsules. It was supposed by this anatomist that each crescent resembled the dentated body in the cerebellum or that in the corpus olivare; but careful examination must convince any one who takes the trouble of it that such is not the fact. It is true that the grey matter contains white fibres, but they mingle with its elements and are not enclosed within a layer of it, as described and delineated by Mayo.

When sections of the spinal cord in different regions are examined, they are found to exhibit differences of dimensions affecting both the white and the grey matter, and remarkable varieties as regards the shape of the lateral portions of the latter. The relative proportion of the grey matter to the white appears to be much greater in the lumbar than in the cervical or dorsal regions. In the upper part of the cord the crescentic portions are narrow, and the white matter is abundant. The posterior horn appears as a thin lamella extending back to the surface, while the anterior is a small, roundish, slightly stellate mass, remote from the surface of the anterior columns. In the dorsal region the grey matter is at its minimum of developement: here it appears much contracted and diminished in size, although presenting the same general form as that in the region of the neck. In the lumbar region both horns acquire a manifest increase of thickness, the posterior still extending back quite to the surface, and the anterior, more stellate than in the higher parts of the cord, separated from the corresponding surface of the cord by a much smaller quantity of white substance. At a still lower part of the cord, where the lumbar swelling begins to diminish in size, the posterior horn is short and thick, and sometimes seems not to reach quite back to the surface of the cord,—an appearance, however, which might be produced by some accidental obliquity of the section; and its posterior extremity has somewhat of the form of a hook, its hindermost portion being directed a little forwards and inwards, forming a very sharp angle with the rest of the grey substance which constitutes the horn. At the lowest part of the cord the crescentic form of the lateral portions of grey matter ceases, and the transverse section of it presents the form of a solid cylinder slightly notched on each side, and surrounded completely by the white substance. (*Fig.* 18.)

There are also differences deserving of notice as regards the white substance in the different regions of the cord. The largest quantity of white substance is found in the cervical enlargement, as may be shown on a transverse section. Both the antero-lateral and the posterior columns are large, but by far the greatest proportion of the mass of white substance must be assigned to the antero-lateral columns. It is also important to remark that the quantity of white substance which is placed between the posterior horns in a great part of the cervical region is augmented by the existence of two small columns of white matter, which will be more particularly described when we come to speak of the medulla oblongata. These columns

SUBSTANCE IN DIFFERENT REGIONS OF THE CORD. 89

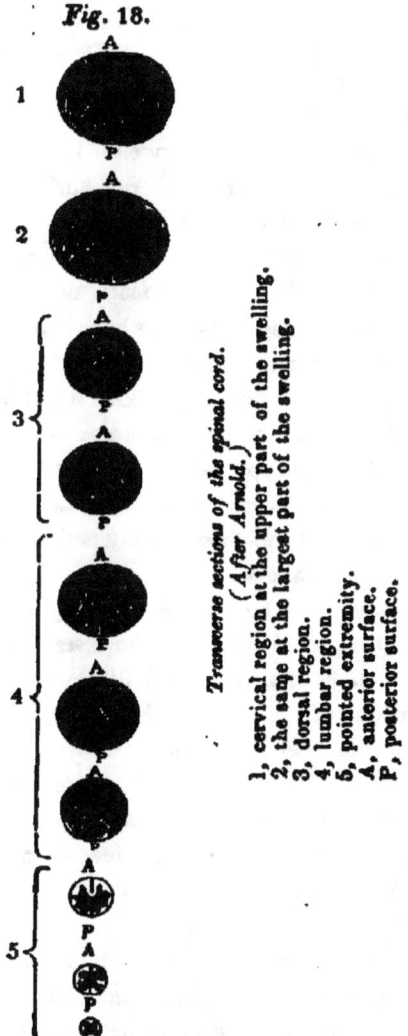

Fig. 18.

Transverse sections of the spinal cord.
(After Arnold.)
1, cervical region at the upper part of the swelling.
2, the same at the largest part of the swelling.
3, dorsal region.
4, lumbar region.
5, pointed extremity.
A, anterior surface.
P, posterior surface.

extend from the inferior extremity of the fourth ventricle, very nearly as far down as the termination of the cervical enlargement, where they gradually taper to a fine point and disappear, allowing the posterior columns of the cord to come into apposition along the posterior fissure. These small columns, *the posterior pyramids* of some authors, do not appear to be completely isolated from the proper posterior columns of the cord. There is generally a very clear line of demarcation between them, visible on the posterior surface, by a distinct depression or fissure which passes in the length of the cord; but this fissure does not extend much deeper than the surface, nor does any distinct process of pia mater sink into it. Nevertheless, in the spinal cord, which has been hardened in alcohol or by any other chemical reagent, these columns will readily separate by tearing in the longitudinal direction, both from each other and from the posterior columns between which they are placed. They occupy rather less than one-half of the interval between the posterior roots of the nerves, excepting at their lowest part, where, from their tapering form, they obviously take up much less space.

In the dorsal region both white and grey matter are small in quantity; the posterior columns, however, do not appear to experience a diminution in size at all commensurate with the general shrinking of the organ in this region, nor with the reduced size of the antero-lateral columns.

In the lumbar region the antero-lateral columns are small; the grey matter is large in quantity, and the posterior columns appear to retain their size: they are, indeed, proportionally to the other parts of the cord, larger in this than in the cervical region, and the lumbar swelling appears to depend much more upon the large size of the grey matter and of the posterior columns than upon the bulk of the antero-lateral ones. At the lowest point of the cord the white matter has gradually disappeared, and in the commencement of the terminal filiform process grey matter only is present, according to Remak and Valentin.

These facts lead to some interesting physiological conclusions bearing upon the function of the cord as well as of its columns. It is in the upper extremities that voluntary power and sensibility are in their most highly developed state, and accordingly the size of that portion of the cord from which the nerves of these parts emanate is greater than any other portion of the cord. Were the sensibility dependent on the grey matter or upon the posterior columns, as has been conjectured, it might most legitimately be expected that a proportionate developement of these parts would exist in the cervical region. Yet a comparison of the cervical with the lumbar swelling demonstrates that the developement of both the grey matter and the posterior columns, (if not absolutely, certainly relatively to the bulk of the segment,) is inferior in the former to that in the latter, whence nerves are supplied to the inferior extremities in which sensibility is much less acute, and in which there is a much less perfect adjustment of the voluntary power to the muscular movements.

The difference of the respective sizes of the antero-lateral columns in those parts of the cord which supply the upper and lower extremities is perfectly consistent with the difference in the sensibility and voluntary power of those parts.[*] And as in the trunk these endowments are at their lowest point of developement, so the dorsal region of the cord is that which exhibits the antero-lateral columns of the smallest bulk.

In the lower parts of the body, which receive their supply of nerves from the lumbar swelling of the cord, there are certain peculiarities worthy of the attention of the physiologist. Thus the sphincter muscles of both the bladder and rectum

[*] Weber's experiments sufficiently indicate that the general as well as the tactile sensibility of the lower extremities is considerably inferior to those of the upper extremities.

are to a great degree independent of voluntary influence, and act independently of consciousness. The principal function of the lower extremities is that of locomotion; they are the pillars of support to the trunk, and the chief agents in the maintenance of its attitudes. And, although in these actions the will exercises a not inconsiderable control, still the principle of purely physical nervous action renders them in a great degree independent of the mind. The reflex or excito-motory actions are much more evident in the lower than in the upper extremities; the former are much more independent of cerebral lesion than the latter. And let it be remarked that these phenomena are associated with high developement of grey matter, and with posterior columns of large size, while the antero-lateral columns are comparatively small. May not the high developement of the grey matter have reference to the exalted state of the physical nervous actions of the lower part of the body, and that of the posterior columns to the locomotive actions? To these points we shall have again to refer when we discuss the functions of the spinal cord.

Is there a central canal in the spinal cord? Many anatomists have affirmed that the spinal cord was traversed in its entire length by a canal, which was continuous with the fourth ventricle. If such a canal exist, it must be extremely difficult to demonstrate, as I have never, after numberless examinations, been able to see it. In transverse sections of the spinal cord, which have been dried upon glass, there is sometimes an appearance which may be attributed to the presence of a minute canal; but I should be more disposed to ascribe it to the patulous mouth of a bloodvessel which had been divided in making the section, for it is by no means constant even in the different regions of the same spinal cord. The situation which some have assigned to this supposed canal is between the grey and white commissures; but Stilling and Wallack* place it in the grey matter. It is obvious that an artificial separation of these layers, which is easily effected, and more especially while the preparation is being dried, would give rise to the appearance of a canal upon a transverse section. It may be stated, further, that the deepest part of the longitudinal fissure is wider than any other portion of it, and, if cut across, might appear like a canal.

The observations of Tiedemann appear to me to put this question in its true light. I shall, therefore, make the following quotation from his learned work on the anatomy of the fœtal brain, without, however, subscribing to the accuracy of all the statements it contains.

" The spinal marrow," says Tiedemann, " represents a hollow cylinder, the thin walls of which are bent backwards, the posterior part representing a longitudinal opening; for it is hollowed by a groove, termed *the canal of the spinal marrow*. This canal exists through the whole cylinder, and communicates with the calamus scriptorius, with the fourth ventricle, which, strictly speaking, is but a dilatation of it. During the first periods we can, without difficulty, separate the thin and flexed walls of the spinal marrow, and thus expose the canal which they contain. This canal is somewhat broader in those points where the spinal marrow sensibly enlarges exteriorly, as at the origin of the nerves for the pectoral and abdominal extremities. The mechanism of its formation is very simple; the pia mater, acquiring more extent, is folded longitudinally backwards and dips into the substance of the spinal marrow, which, as we have seen, had been previously in a fluid state.

* Untersuchungen über die Textur des Rückenmarks. Leipz. 1842.

It is very evident that, in the commencement of the second, third, and even fourth months, this canal has, in proportion to the thickness of the walls of the spinal marrow, a much greater capacity than it subsequently acquires. The contraction which it undergoes in the progress of the developement of the embryo, arises from the pia mater depositing a new substance, the materials of which it derives from the blood sent by the heart, and which, augmenting the volume of the walls of the cylinder, ought necessarily to diminish the calibre of the central canal. This substance is soft, reddish, and traversed by numerous small vessels during the period of the last two months. We cannot doubt, then, that the grey substance of the spinal marrow has an origin subsequent to that of the medullary fibrous substance, and that it is applied from within outwards on the surface of this latter. Consequently, the opinion of M. Gall, that it is formed prior to the medullary, and is, as it were, the matrix, is absolutely false with regard to the spinal marrow, for we already perceive the roots of the spinal nerves, in the second and third months, although at this period there is no cortical yet deposited in its canal."

" It is very remarkable that the canal of the spinal marrow exists constantly and during the entire life of the animal, in fishes, reptiles, and birds. I have met it in a great number of fishes, both of salt and fresh water, such as the ray (*Raia*), shark (*Squalus*), bream (*Cyprinus brama*), bandfish (*Cepola*), pike (*Esox*), salmon, carp, &c.; and I have always found its internal surface covered with a layer of grey substance. The observations of M. Arsaky agree perfectly with mine in this respect."*

" I have observed the same canal, in question, in the

* Diss. de piscium cerebro et medulla spinali. Halle, 1813, p. 9.

hawk's-bill tortoise, common tortoise, a young crocodile of the Nile, wall lizard, ringed snake, land salamander, green frog, and the common toad. In front it is continuous with the fourth ventricle, or rather it dilates to give origin to this cavity, and its interior was covered with a thin layer of cortical substance."

" Birds possess this canal both in their embryo state and in adult age. In these it forms, at its inferior part, a remarkable excavation, which Steno, Perault, Jacobœus, and some other authors have described under the name of the rhomboidal sinus. In birds, also, the grey substance occupies the interior, and is no where in greater abundance than on the walls of this sinus."

" The canal equally exists in the spinal marrow of the fœtus of mammiferous animals, as also in the young animals of this class (?) F. Meckel has found it in the embryo of the rabbit;* and G. Sewell in young animals of the genus dog, sheep, ox, and horse.† This latter writer observes that it was filled with a colourless fluid, nearly opaque, and of the same nature as that which existed in the ventricles. F. Meckel has even met a small canal full of fluid in the spinal marrow of some of the adult mammiferous class, such as the dog, cat, rabbit, sheep, and ox. Blaes has met it also in many adult mammiferous animals.

" Although we cannot find this canal in the spinal marrow of the human adult in its normal state of developement, still it has undoubtedly been met with; we should, then, consider it as the result of a retardation in its developement. Charles Stephen‡ was the first who gave a description of it; and Co-

* Beiträge zur Vergleichend: Anatomie, cap. ii. No. i. p. 32.
† Phil. Trans. for 1809.
‡ De dissectione partium corporis humani, lib. iii. Par. 1545.

lumbo,* Piccolhomini,† Bauhin,‡ Malpighi,§ Lyser,‖ Golles,¶ Morgagni,** Haller,†† and M. Portal,‡‡ have since observed it. Many of these writers have even considered it as a constant and normal disposition; an hypothesis which Varoli, Monro, Sabatier, and some other anatomists have justly opposed. Nymman proceeded even still further, for he spoke of two canals prolonged into the spinal marrow. Gall pretends to have found in the spinal marrow of new-born infants, in infants of a certain age, and even in certain adults, two canals free from all communication with the fourth ventricle, but which extended through the pons Varolii, the tubercula quadrigemina and the medulla oblongata into the interior of the optic thalami, where they formed a cavity sufficient to lodge an almond! These two supposed canals, with their termination in the optic thalami, do not exist; we must suppose that they are produced by a forced insufflation: I have never met them either in the adult or in the fœtus; nor do we find them in animals in which the canal of the spinal marrow always communicates with the fourth ventricle, by means of the calamus scriptorius."§§

* De re anatomicâ, Ven. 1559.
† Anatom. Prælectiones, Rom. 1586.
‡ Theatrum Anatomicum. Francf. 1605.
§ De cerebro, in his Opera Minora, t. ii. p. 119.
‖ Culter Anatomicus. Copenhag. 1653.
¶ Abrégé de l'œconomie du grand et petit monde. 1670.
** Adversar. Anatom. Animadv. xiv.
†† Elem. Physiologiæ, t. iv.
‡‡ Observ. sur un spina bifida et sur le canal de la möelle epiniere; dans Mém. de l'Acad. des Sc. 1770.
§§ Dr. Bennett's translation of Tiedemann's Anatomy of the Fœtal Brain, pp. 124 et sqq.

Bloodvessels of the spinal cord.—The arteries of the spinal cord are derived from the vertebral arteries as well as from the small vessels which ramify upon the spinal column in the cervical, dorsal, and lumbar regions.

Of these the largest and most important are the two spinal arteries which spring from the vertebral on each side, distinguished as the *anterior* and *posterior* spinal arteries.

The *anterior* spinal artery is the larger of the two. It arises from the vertebral artery near to the basilar: sometimes it comes from the basilar itself, or from the inferior cerebellar; and sometimes the arteries of opposite sides have different origins, one arising from the vertebral and the other from the basilar. It passes nearly vertically downwards, inclining inwards, in front of the medulla oblongata, and having passed for a short distance in front of the cord, it unites at an acute angle with its fellow of the opposite side, forming a single vessel, which passes down in front of the anterior median fissure, under cover of the band of white fibrous tissue which is found along the middle of the anterior surface of the cord. The artery thus formed is called the *anterior median artery* of the spinal cord.

" The anterior or median spinal artery," says Cruveilhier, " therefore, results from the anastomoses of the two anterior spinal branches of the vertebral. In one case there was no artery on the left side, but the right was twice as large as usual. The vessel is of considerable size until it has passed below the cervical enlargement of the cord, from which point down nearly to the lumbar enlargement it becomes exceedingly delicate; a little above the last-named enlargement it suddenly increases in size, again gradually diminishes as it approaches the lower end of the spinal cord, and becoming capillary is prolonged down to the sacrum, together with the fibrous string in which the spinal cord terminates."

" During its course this artery receives lateral branches from the ascending cervical and the vertebral in the neck, and from the spinal branches of the intercostal and lumbar arteries in the back and loins. Those branches penetrate the fibrous canal formed by the dura mater around each of the spinal nerves; become applied to the nervous ganglia to which they supply branches, yet intermixed with and follow the course of the corresponding nerves; send small twigs backwards to the posterior spinal arteries, and terminate in the anterior spinal trunk at variable angles, similar to those at which the nerves are attached to the cord." *

The posterior spinal arteries arise from the vertebral or from the inferior cerebellar artery: they incline backwards to the posterior surface of the spinal marrow, along which they descend in a tortuous manner, anastomosing freely with each other and with the small arteries which accompany the nerves in the intercostal foramina. A network of vessels surrounds each posterior root of a spinal nerve, derived from ramifications of those arteries. We can trace the posterior spinal arteries as low down as the lumbar region, distinct throughout their entire course.

Veins.—The blood is returned from the spinal cord by a venous plexus which emerges from the pia mater and is spread over its whole surface: opposite the roots of each nerve a small vein is formed, which passes outwards with the nerve in the same sheath, and empties itself into the large vein which is situate in the intervertebral foramen. Veins accompany the anterior and posterior spinal arteries in the upper part of their course. Branches from this plexus frequently pass to the dura mater involved in a fold of arachnoid, and thus communicate with the general plexus which surrounds the sheath.

* Cruveilhier, Anat. Descr.

We observe that the arteries of the spinal cord are reduced to a very minute size before they penetrate the substance of that organ. The largest vessels are therefore found on its surface or in its fissures. And it may be further remarked that when vessels of a size to be readily detected by the naked eye penetrate the substance, numerous foramina, produced by the separation of the nervous fibres, become distinctly visible. This is very obvious in the white commissure.

The purpose of such a minute subdivision of bloodvessels, prior to their entrance into the substance of the cord, must evidently be to guard the nervous substance against the impulse of several columns of blood of large size. A similar provision, made in a more conspicuous manner, is manifest in the brain, and will be noticed by-and-bye.

Of the spinal nerves.—There is a pair of spinal nerves for each intercostal foramen, and for that between the atlas and occiput. We can thus enumerate in all thirty-one pair of nerves having their origin from the spinal cord, and this number is exclusive of the spinal accessory nerve which is connected with the upper part of the cervical region.

The spinal nerves have the following very constant characters. Each has its origin by two roots, of which the anterior is distinctly inferior in size to the posterior. The ligamentum denticulatum is placed between these roots. Each root passes out through a distinct opening in the dura mater. Immediately after its emergence a ganglion is formed on each posterior root, and the anterior root lies embedded in the anterior surface of the ganglion and involved in the same sheath (*fig.* 19), but without mingling its fibres with those of the ganglion. Beyond it, the nervous fibres of both roots intermingle, and a compound spinal nerve results. The trunk thus formed passes immediately through the intervertebral tube and divides into an anterior and posterior branch, which are

THE SPINAL NERVES.

Fig. 19.

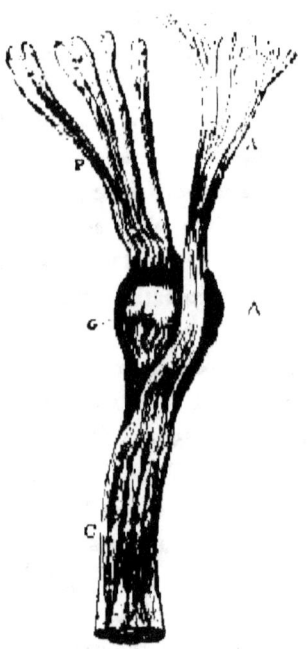

Origin of a spinal nerve.
(After Bell.)

A, A, anterior root.
P, posterior ditto.
G, ganglion on the posterior root.
C, compound nerve resulting from the commingling of the fibres of both roots.

distributed to the muscles and integument of the trunk and the extremities. Of these branches the anterior one is generally much the larger.

An exception, however, to this arrangement occurs in the case of the first spinal nerve (the tenth pair of Willis), to which Winslow gave the appropriate name *sub-occipital nerve*, to indicate its peculiarity of character. This nerve sometimes has only one root, and that corresponds to the anterior. More generally it has two roots, of which, unlike the other spinal nerves, the anterior is the larger, containing, according to Asch, from three to five or seven bundles of filaments, whilst the posterior contains two or three, or at most four much smaller bundles. Very frequently the posterior filaments of either the right or left side unite with the spinal accessory, a slight enlargement or knot being formed at the point of junction; from this place a bundle of filaments emerges equal in size to the posterior root, and takes the ordinary course of that root, a small ganglion being formed upon it at the usual situation. Frequently, however, this ganglion is wanting. The compound nerve formed from the junction of these two roots, besides giving off communicating filaments to the sympathetic, divides as the other spinal nerves do into an anterior and posterior root, of which, however, contrary to the usual arrangement, the posterior is the larger.

The spinal nerves are arranged naturally into classes according to the regions of the spine in which they take their rise. We number eight in the cervical region, the sub-occipital included; twelve in the dorsal region; five in the lumbar, and six in the sacral regions. All the nerves, after the second, pass obliquely outwards and downwards from their emergence from the spinal cord to their exit from the vertebral canal, and this obliquity gradually increases from the higher to the lowest nerves. The roots of the nerves possess certain characters, of which some are common to all, and others are peculiar to the nerves of particular regions.

All the spinal nerves arise from the cord by separate fasciculi of filaments, which, as they approach the dura mater, converge to each other and are united together to constitute the anterior or the posterior roots. The posterior roots of opposite sides lie at a pretty uniform interval, from the upper to the lower part of the cord, indicating but a very trifling change in the thickness of the posterior columns throughout their entire course. The ganglia on the posterior roots are all proportionate to the size of their respective roots.

The characters proper to the nerves of particular regions may be stated as follows:*—

The cervical nerves exhibit much less obliquity of their roots than the other vertebral nerves. The second cervical nerve is transverse (the first passing a little upwards as well as outwards); the succeeding nerves slope downwards and outwards, the lowest being the most oblique; the obliquity, however, never exceeds the depth of a single vertebra.

The roots of the nerves in the cervical region are of considerable size. The posterior roots bear a larger proportion to the anterior than in any other part of the spine. According to Cruveilhier, the ratio is as 3 to 1, and this estimate is probably correct. It applies not only to the entire root, but to the fasciculi of filaments which enter into their formation.

The nerves in this region increase rapidly from the first to the fifth, and then maintain nearly the same size to the eighth.

The dorsal nerves, with the exception of the first, which closely resembles a cervical nerve, have very peculiar characters.

There is a manifest increase in the obliquity of the roots, the length of each root within the spinal canal being equal the height of at least two vertebræ. And it may be remarked that the apparent obliquity is less than the real, for each root remains in contact with the cord for a short distance after its actual emergence from the substance of it, and consequently the point of separation is some way below the point of emergence.

---

* In the succeeding statements I have followed Cruveilhier's description, which I have verified, excepting in a few points, which are specified.

The interval between the roots is greater in the dorsal region than any other segment of the cord. The bundles, which compose the roots, are smaller than elsewhere.

We observe a very slight disproportion between the anterior and posterior roots in the dorsal region. The latter, however, still maintain predominance of size.

Lumbar region.—From the dorsal region to its terminal extremity the surface of the spinal cord is covered both on its anterior and posterior aspects by the fasciculi of origin of the lumbar and sacral nerves. They emerge very close to each other upon those surfaces, and the intervals between the sets of fascicles proper to each root are extremely short, so that they form an uninterrupted series of bundles on each surface.

The proportion of the posterior to the anterior roots in the lumbar region is as 2 to 1 according to Cruveilhier, or as 1½ to 1, which seems to me to be nearer the truth. And there does not appear to be any material difference in point of size between the posterior and anterior fascicles.[*]

A very interesting feature in the origin of the lumbar and sacral nerves is apparent from observing the relation to each other of the anterior and posterior roots of opposite sides on the respective surfaces of the cord. The anterior roots of opposite sides may be seen to approximate the median line gradually as they descend, until at the lowest points they almost touch. On the contrary, the posterior roots continue nearly in the same sequence all the way down. It may, therefore, be supposed that in the tapering of the cord the anterior fibres of the antero-lateral columns separate most quickly from it.

[*] M. Blaudin assigns the following proportions of the posterior to the anterior roots in the several regions; in the cervical region as 2 : 1; in the dorsal region as 1 : 1; and in the lumbar and sacral regions as 1½ : 1.

The direction of these roots is almost vertical, and their length within the canal of the dura mater is very considerable. The aggregate of them forms the *cauda equina*.

I have not observed that the situation in which the ganglia of the sacral nerves are formed is different from others. They are contained, as elsewhere, in sheaths of dura mater, and lie in the sacral foramina, surrounded by fat, and from the looseness of their connection with the walls of those foramina they may be very easily detached.

I cannot confirm Cruveilhier's statement that the anterior and posterior roots in the sacral region together form the ganglia.

The roots of the sacral nerves gradually diminish in size, so that the lowest are smaller than any others which emerge from the spinal cord.

In connecting the peculiar anatomical characters of the spinal nerves in the various regions with their physiological action, some interesting points are presented to our notice.

The great size of the cervical nerves is quite in conformity with the exalted vital actions of the upper extremities. And the predominance of the posterior over the anterior roots, both positive, and as compared with other regions, corresponds with the great developement of sensation in the upper limbs.

The posterior root of the second cervical nerve, as has been noticed by Longet, is considerably larger than the anterior, as 3 : 1; and it is from this source that the occipital and mastoid nerves, the sensitive nerves of the integument in the occipital region, derive their filaments.

In the dorsal region the almost equality of the anterior and posterior roots and the small size of both is consistent with the absence of any great degree of developement either of the sensitive or motor power.

Lastly, the increased muscular activity of the lower extremities and their greater sensibility as compared with the dorsal segment, renders necessary the increase of size which the roots of the lumbar and sacral nerves experience. And it may be conjectured that the predominance of the posterior roots has reference to the exalted sensibility of some parts of the lower limbs.

One of the most important problems in the anatomy of the spinal cord is to determine the precise relation which the roots of the nerves bear to the columns of the cord and to the grey matter. As far as coarse dissection enables me to determine, I would venture to make the following statement, founded upon my own observations.

The anterior roots derive their fibres wholly from the antero-lateral columns. Of these fibres it is probable that some are continuous with the longitudinal or oblique fibres of the cord, and that others pass into the grey matter. This, however, is very difficult, if it be possible, of demonstration by the ordinary modes of dissection. The posterior roots adhere to the posterior part of the antero-lateral columns, and derive their fibres chiefly from that source. I have never, in numerous dissections, seen any thing to induce me to believe that the posterior columns contribute to the formation of the posterior roots. If they do, it must be by few and extremely delicate fibres. It seems highly probable (although the demonstration of the fact is attended with great difficulty) that the fibres of the posterior roots have a similar disposition to that described for the anterior, and that some pass into the posterior horn of the grey matter, and others are continuous with the longitudinal fibres.

Various conflicting statements have been made by the anatomists who have written upon the spinal cord, with regard to the actual connection of the roots of the nerves with the proper substance of the organ. Nor is this to be wondered at, when we consider the great delicacy of the investigation. It is very easy to trace any set of filaments to the pia mater; but after they have passed beyond that covering, the nervous fibres lose their main support and their bond of union, and they separate from each other. Their exquisite delicacy and microscopic size render any further dissection of them extremely difficult. Mr. Grainger, in his excellent treatise on the spinal cord,* recommends certain precautions which I have adopted with advantage. The cord should be examined *immediately after death*, as the delay even of a few hours increases the softness of the medullary substance. Great advantage is derived from placing the cord, immediately after its removal, in a very weak mixture of alcohol and water, as by these means firmness is given to the parts without rendering them crisp and brittle, as happens if strong alcohol be used. The parts should be dissected with very fine instruments under water. " I have met with most success," says Mr. Grainger, " by dividing the pia mater at the median fissure, and very cautiously raising it as far as the lateral furrow, leaving its connection with the fibres of the nerves intact; it is then necessary to open either the anterior or posterior lateral fissure, according to the root examined, at a little distance above the exact place where the nerve which is to be dissected is attached to the cord, when by cautiously proceeding to open the fissure, the threads which dip into the grey matter are perceived." Mr. Grainger recommends the adoption of a similar mode of dissection for the cranial nerves, care being taken in every case not to disturb the connection of the pia mater with the nervous fibres themselves. He also very properly cautions the dissector against a deceptive appearance connected with the passage of those bloodvessels which enter the lateral fissure, in order to reach the internal grey substance. " Without due

* Observations on the Structure and Functions of the Spinal Cord, p. 37. Lond. 1837.

precaution," he adds, " these vascular branches may themselves be readily mistaken for nervous fibrils; but they are especially liable to be productive of error, because, when they are made tense, they cause those portions of the longitudinal fibres of the cord, which are left between them, to assume exactly the appearance of flat transverse fibres; this circumstance probably misled Gall, and induced him to suppose that all the fibres of the spinal nerves were connected with the grey substance."

The following is Mr. Grainger's account of the result of his examinations conducted with the precautions above specified.

" After repeated examinations, I satisfied myself that each root was connected both with the external fibrous part of the cord and the internal grey substance. The following is what appears to be the structure : after the two roots have perforated the theca vertebralis, and so reached the surface of the cord, it is well known that their fibres begin to separate from each other; of these fibres, some are lost in the white substance, whilst others, entering more deeply into the lateral furrows, are found to continue their course, nearly at a right angle with the spinal cord itself, as far as the grey substance in which they are lost. But this arrangement has no resemblance to the distinct division into fasciculi, depicted by Mr. Mayo; on the contrary, it is with great care only that small, delicate, individual threads or striæ, as it were, are traced, dipping into the lateral fissure, and at length joining the grey matter. This difficulty is owing to the fact that whilst the fibres on the outer surface of the pia mater adhere very intimately with that strong membrane, on its inner surface, the membrane becomes so extremely delicate, that the fibres lose much of their firmness, and break on the application of the least force; an accident which always happens if the pia mater be raised from the surface of the spinal cord, beyond the point where the nerves are attached. When the filaments have penetrated into the fissure, they lose their rounded figure and become flattened, and are then seen passing to the grey substance at a right angle to the longitudinal fibres. It is extremely difficult, owing to the delicacy of the parts, to determine the exact relations which exist between the above filaments and the grey matter; but in a few dissections I have been able to perceive these fibrils running like delicate striæ in the grey substance. In one instance the fibres, being more distinct than usual, an appearance was presented having a remarkable resemblance to that which is seen on making a section of the corpus striatum in a recent brain, after the method of Spurzheim. My friend and colleague, Mr. Cooper, in this case counted distinctly five separate fibrils passing from the anterior root of one nerve, and there were some other fibres derived from the same root, which were not so plainly seen."

" From numerous examinations," continues Mr. Grainger, " I am induced to believe that whenever the white fibres of the nervous system become connected with the grey substance, whether in the different masses of the brain, in the spinal cord, or in the ganglions, the arrangement is similar to what is seen in the section of the corpus striatum to which reference has just been made. The fibres become as it were encrusted with the grey matter, a disposition which may even be seen by a careful inspection in the convolutions of the cerebrum, in which the radiating fibres of the crus cerebri are observed like delicate striæ."

I have repeated the dissections of the roots of the nerves in the manner described by Mr. Grainger, and am enabled to confirm his general results. It appeared to me, however, that considerably the greater number of the fibres passed in at right angles, whilst those which might be supposed to take an upward course were few and indistinct, and seemed rather to pass obliquely inwards and slightly upwards than to assume the vertical direction. In short, when the fibres had penetrated the medullary substance, they seemed to diverge from one another,—those which occupied a central position preserving much more of parallelism than either the upper or the lower ones. It is extremely difficult to demonstrate the direct continuity between the fibres of the nervous roots and those of the cord. Valentin has, indeed, depicted the transition of nerve fibres into the spinal cord as seen by the microscope; but these may be passing to the grey matter of the cord. The continuity of the fibres of the nerves with the longitudinal fibres of the cord would probably take place at the surface of the latter in greater numbers than at more deeply-seated planes. In the dissections above described, such fibres would be very apt to be destroyed or to be overlooked. Mr. Grainger, in the work before referred to, speaks evidently with much greater confidence of the connexion of the roots of the nerves with the grey matter than of their continuity with the longitudinal fibres. He expresses his conviction, however, that such a continuity does exist, although the exact mode of connexion and the situation at which it occurs cannot be demonstrated.

This question, respecting the precise relation of the roots of the nerves to the cord, is one of those in which physiology in a certain sense takes the lead of anatomy. Experiment has made it certain that while the spinal cord serves as a propagator of nervous power to and from the brain, as in the ordinary sensations and voluntary movements of the trunk and extremities, it is likewise capable of acting as an independent nervous centre, and that movements of a very definite character may be produced in parts connected with it, even after all communication between it and the brain has been cut off. And it has been supposed by one of the most zealous labourers in this department of physiology, that a distinct series of nervous fibres is directed to each class of actions, those, namely, of sensation and volition, and those which are independent of the brain. Mr. Grainger was the first who offered a distinct solution to the anatomical problem which arose out of this hypothesis. Probable as his explanation appears to be, a candid review of the observations which have been hitherto made obliges me to state my opinion that the question is still *sub judice*, and that further research is necessary to prove unequivocally that, of the fibres composing the roots of the nerves, some pass upwards and enter the brain, and others do not pass beyond the grey matter of the spinal cord. And this inquiry demands more than ordinary care, for the mind of an observer would be easily biased by so attractive an hypothesis as that above referred to.

It is not from physiological experiment nor from coarse dissections that we can expect a solution of this difficult but most important problem. We must look to the microscopical analysis of the anatomical elements of the spinal cord, as well as of the encephalon, for the most exact results upon all questions connected with the working of these centres. In a subsequent part of this volume I shall give an account of the present state of our knowledge of this most interesting subject, having first examined the coarser anatomy of the several parts of the encephalon.

CHAPTER IV.

The Encephalon—Its weight, absolute and relative—Tiedemann's observations—The brain in different races of mankind—Methods of examining the encephalon—Methods of Willis, Reil, Gall and Spurzheim—Surface of the encephalon—Dissection of the brain from above.

THE term *Encephalon* is used here in its strictly etymological sense to denote that part of the cerebro-spinal centre which is contained within the cavity of the cranium. Although it forms a great mass, continuous throughout, it offers certain very obvious subdivisions, which may be more conveniently described separately. Before proceeding, however, to the description of these portions, it will be necessary to take a brief review of some general points connected with the entire encephalic mass.

The size of the encephalon by no means keeps pace with that of the body. In comparing that of the four classes of vertebrate animals, we observe a manifest increase of its size as compared with the body in the following order, fishes (minimum), reptiles, birds, mammalia. This statement, although applicable to the encephalic mass when viewed as a whole, does not apply to certain of its parts, which are often more developed in the less perfect than in the more highly organized animals. The cerebrum and cerebellum, however, exhibit this gradual increase of developement, and their enlargement is in accordance with a gradually increasing manifestation of mental faculties. And it is upon the great size of these parts that the superiority of the human brain over that of all other animals depends.

In comparing the brains of some of the larger mammalia with that of man, we observe an evident want of correspondence between the bulk of the encephalic nerves and that of the encephalon itself. This does not accord with what we have had occasion to notice respecting the spinal cord, in which large nerves were always concomitant with high developement of the organ itself. The maximum weight, as Müller remarks, of a horse's brain is, according to Soemmering, 1 lb. 7 oz.; the minimum of an adult human brain 2 lb. 5½ oz.; nevertheless, the nerves at the base of the brain are ten times thicker in the horse than in the human subject.

This want of correspondence between the developement of the mass of the body and that of the brain, as well as between the size of that organ and of the encephalic nerves, must surely be admitted to indicate an incorrectness in the assertion of the distinguished physiologist who has just been quoted, namely, that "all the primitive fibres of the nerves terminate in the brain; those of the cerebral nerves immediately, those of the spinal nerves through the medium of the spinal cord."[e] The human brain must evidently contain numerous other fibres besides those which are continuous with the roots of the nerves, and it is likely that the horse's brain contains similar ones, although less numerous; it seems, therefore, impossible that the small brain of the horse can be the point of convergence of the large spinal and cerebral nerves of that animal; and if this be true as regards the horse, it is so likewise in man. It is much more probable that a large proportion of them do not extend beyond the spinal cord, and that the greater number of

[e] Physiol. transl. by Baly, 2nd ed. p. 796.

the fibres of the encephalic nerves do not go beyond that segment of the encephalon in which they are immediately implanted.

It must be admitted, however, that although this disproportion is very manifest as regards the whole encephalon, it is not so evident when we compare the nerves with those segments of the organ from which they immediately arise. Thus, the medulla oblongata is always, as regards mere bulk, in the direct ratio of its nerves; the optic lobes are large when the optic nerves are so; the olfactory lobes bear a close relation to the number of the olfactory nerves, and it may be added, to the complication of the olfactory organ. It is to the cerebral hemispheres, to the cerebellum and the systems of fibres immediately connected with them, that we must attribute the disproportion in question: those parts being small when the nerves are large, as in the horse, and large when the nerves are of small size, as in man.

The human encephalon weighs about 48 oz. for the male, and 44 oz. for the female.* This estimate, which was formed by Krause, does not differ very materially from that derived from Professor John Reid's careful observations made at the Royal Infirmary at Edinburgh. The following tables are extracted from a paper by this excellent anatomist in the London and Edinburgh Monthly Journal of Medical Science for April, 1843.†

---

* According to Mr. Hamilton's investigations, the adult male brain in the Scot's head weighs, on an average, 3 lbs. 8 oz. troy; about one brain of seven is found about 4 lbs. troy; the female brain weighs 3 lbs. 4 oz.; and one of a hundred female brains weighs 4 lbs.

† Reference may also be made to an extensive series of observations on the weight of the brain by Dr. Sims, Med. Chir. Trans. vol. xix. p. 353; also to some by Dr. Boyd, quoted in Dr. Willis's edition of Wagner's Physiology; and to a paper by M. Parchappe, Gazette Médicale, Oct. 8, 1842.

Table I.

Average weight of the encephalon, &c. between 25 and 55 years of age, in the two sexes, and the average difference between them—Males, 53 brains weighed—Females, 34 brains weighed :—

	Male.			Female.			Difference in favour of Male.	
	lb.	oz.	dr.	lb.	oz.	dr.	oz.	dr.
Average weight of encephalon		50	$3\frac{1}{3}$ or		44	$8\frac{1}{2}$ or	5	11
	3	2	$3\frac{1}{3}$	2	12	$8\frac{1}{2}$		
Cerebrum		43	$15\frac{3}{4}$		38	12	5	$3\frac{3}{4}$
Cerebellum		5	4		4	$12\frac{1}{4}$	0	$7\frac{3}{4}$
Cerebellum, with pons and medulla oblongata		6	$3\frac{3}{8}$		5	$12\frac{1}{3}$	0	$7\frac{1}{2}$ nearly

Table II.

Relative weight of encephalon to cerebellum, and to cerebellum with medulla oblongata and pons Varolii, between 25 and 55 years of age, in the two sexes (53 male and 34 female brains weighed).

	Male.	Female.
Relative weight of encephalon to cerebellum	as 1 to $9\frac{4}{7}$	as 1 to $9\frac{1}{4}$
Ditto to ditto, with pons and medulla oblongata	1 „ $8\frac{8}{13}$	1 „ $7\frac{9}{16}$

From this table it would appear that, in the female, the average cerebellum is, relative to the encephalon, a little heavier than in the male.

In a third table, which has been reduced from that published by Professor Reid, the average weight of the encephalon, cerebellum, with pons Varolii and medulla oblongata, is given over a much wider range of age than that in the first table quoted. For this purpose 253 brains were weighed.

TABLE III.

MALES.

Age. Years.	Number weighed.	Encephalon. oz. dr.	Cerebellum. oz. dr.	Cerebellum with pons & medulla ob. oz. dr.
1 to 4	5	39 .. 4⅛	3 .. 13¼	4 .. 6⅞
5 — 7	3	43 .. 10	4 .. 7	5 .. 6
7 — 10	6	46 .. 2⅜	4 .. 10⅔	5 .. 10⅔
10 — 13	3	48 .. 7½	4 .. 14	5 .. 12
13 — 16	5	47 .. 8⅔	—	6 .. 1½
16 — 20	6	52 .. 10	5 .. 4¼	6 .. 6¼
20 — 30	25	50 .. 9¾	5 .. 3¾	6 .. 2
30 — 40	23	51 .. 15	5 .. 3¾	6 .. 4½
40 — 50	34	48 .. 13½	5 .. 3¾	6 .. 4$\frac{4}{11}$
50 — 60	29	50 .. 2	5 .. 5$\frac{5}{18}$	6 .. 2$\frac{2}{14}$
60 — 70	8	50 .. 6⅝	5 .. 0	6 .. 2
70 & upw.	7	48 .. 4⅞	4 .. 14	5 .. 14⅞

Total male brains weighed 154.

FEMALES.

Age. Years.	Number weighed.	Encephalon. oz. dr.	Cerebellum. oz. dr.	Cerebellum with pons & medulla ob. oz. dr.
2 to 4	6	37 .. 9	3 .. 9¼	4 .. 5
5 .. 7	3	39 .. 9¾	3 .. 11	4 .. 8⅞
7 .. 8	3	42 .. 7½	4 .. 7¼	5 .. 5½
16 .. 20	8	44 .. 11½	4 .. 14¼	5 .. 11
20 .. 30	18	45 .. 2⅞	4 .. 11¼	5 .. 9¼
30 .. 40	23	44 .. 1½	4 .. 13¼	5 .. 11
40 .. 50	18	44 .. 10⅔	4 .. 14	5 .. 14¼
50 .. 60	5	45 .. 4¼	4 .. 7½	5 .. 8¾
60 .. 70	11	42 .. 14⅞	4 .. 10$\frac{7}{11}$	5 .. 9
70 & upw.	2	38 .. 8½	4 .. 5½	5 .. 2½

From this table we are led to conclude that the brain reaches its greatest absolute weight at an early age. The maximum is found in the table at between 16 and 20; but, as Dr. Reid states, it is plain that the apparent excess of weight at this period over that for the next forty years must have arisen from sources of fallacy incidental to insufficient data. And in the group between 40 and 50, Dr. Reid states that some brains much below the average weight were found, so as to leave no doubt that the diminution in the average weight in that group was attributable to that circumstance.

A decided diminution in the average weight of the brain was noticed in females above 60 years of age; but, among the males, this was not apparent until a later period. Upon this point Professor Reid makes the following judicious observation, which I am anxious to quote as according with the views I have expressed at pp. 48, 49, respecting liquid effusions. " We certainly did expect," he says, " also to find a similar diminution in the average weight of the male brain above 60 years of age, for we are perfectly satisfied, as the tables containing the individual facts will shew, that we more frequently meet with a greater quantity of serum under the arachnoid and in the lateral ventricles in old people than in those in the prime of life. We are also satisfied from an examination of the notes we have taken at the time the brains were examined, that a certain degree of atrophy of the convolutions of the brain over the anterior lobes, marked by the greater width of the sulci, was more common in old than in young persons. We have, however, frequently remarked these appearances in the brains of people in the prime of life who had been for some time addicted to excessive indulgence in ardent spirits."

The ratio between the weight of the body and that of the brain is greater in early age than at the subsequent periods. The following proportions were obtained by Tiedemann, in infants just born. In two boys the proportion of the brain to the body was as 1 : 5.15 and 1 : 6.63, and in two girls as 1 : 6.29 and 1 : 6.83.

The general conclusions deducible from the preceding statements are, that the human brain reaches and maintains its highest degree of developement between the ages of 20 and 60; that the female brain is materially smaller than that of the male; that the proportion of the weight of the brain to that of the body decreases with age, and the most marked diminution in this respect takes place between the ages of 20 and 30 years, although it has already begun at 5 years, and occurs very decidedly at from 13 to 15 years; and lastly, that the great preponderance of the human brain over that of most of the lower animals depends upon the great developement of the cerebrum and cerebellum.

It was formerly admitted, pretty generally, that the human brain was larger, both absolutely and relatively to the size of the body, than that of any other animal. This assertion, however, must now be received with some modification. Exceptions to its superiority in absolute weight are found in the elephant and the whale. The brain of an African elephant, seventeen years old, which was dissected by Perrault, weighed 9 lbs.* The brain of an Asiatic elephant weighed, according to Allen Moulins, 10 lbs.† Sir Astley Cooper dissected an elephant's brain, which weighed 8 lbs. 1 oz. 2 grs. (avoirdupois.)‡ Rudolphi found that the brain of a whale, 75 feet long, (*Balæna mysticetus,*) weighed 5 lbs. 10½ oz., and that that of a narwhal (*Monodon monoceros,*) 17 to 18 feet long, had a weight of 2 lbs. 3 oz. And there are likewise exceptions to the statement that the human brain is larger than that of other animals, relatively to the size of his body. Pozzi§ has shewn (as quoted by Tiedemann) that many small birds (for instance, the sparrow) have, in comparison to the size of their body, a larger brain than man; and Daubenton, Haller, Blumenbach, and Cuvier, found the brain of some of the smaller apes, of the

* Descr. Anatom. d'un Elephant, Mém. de l'Acad. des Sciences de Paris, t. iii.

† An anatomical account of an Elephant. Lond. 1682.

‡ Quoted in Tiedemann's paper on the Brain of the Negro. Phil. Trans. 1836.

§ Observat. Anatom. de Cerebro, an sit in homine proportione majus, quam in aliis animalibus?

Rodentia, and singing birds, relatively to the size of the body, larger than in man.*

"We must seek for the cause of man's superiority," says Tiedemann, "not merely in the greater bulk of his brain in comparison to that of his body, but regard must also be had to the size of his brain with respect to the bulk and thickness of his cerebral nerves, and likewise to the degree of perfection in its structure. Soemmering was the first to show that the human brain, in comparison to the size and thickness of the nerves, is larger than that of any other animal, even the elephant and whale, both of which have an absolutely larger brain than man. Blumenbach's, Obels', Cuvier's, Treviranus', and my own researches have sufficiently corroborated this. It is also satisfactorily shewn that the organization of the human brain is far superior to that of any other animal, not even excepting those apes which bear the closest resemblance to man."

The following conclusions, which Tiedemann deduces from his observations, are so important that I cannot refrain from inserting them here.†

"1. The weight of the brain of an adult male European varies between 3 lbs. 2 oz. and 4 lbs. 6 oz. The brain of men who have distinguished themselves by their great talents is often very large. The brain of the celebrated Cuvier weighed 3 lbs. 11 oz. 4 dr. 40 grs. avoirdupois, or 4 lbs. 11 oz. 4 dr. 30 grs. troy weight. The brain of the celebrated surgeon Dupuytren weighed 4 lbs. 10 oz. troy weight. (Both of these eminent individuals, it ought to be remarked, died with the brain in a state of disease.) The brain of men, with feeble

---

* Tiedemann's paper on the Brain of the Negro, before quoted. See also Leuret's Table, Anat. Comp. du Systeme Nerveux, t. i. p. 420.

† Loc. cit. p. 502.

intellectual powers, is, on the contrary, often very small, particularly in congenital idiotismus. The brain of an idiot, fifty years old, weighed but 1 lb. 8 oz. 4 dr., and that of another, forty years of age, weighed but 1 lb. 11 oz. 4 dr.

" 2. The female brain is lighter than that of the male. It varies between 2 lbs. 8 oz. and 3 lbs. 11 oz. troy. I never found a female brain that weighed 4 lbs. The brain of a girl, an idiot, sixteen years old, weighed only 1 lb. 6 oz. 1 dr. The female brain weighs, on an average, from four to eight ounces less than that of the male; and this difference is already perceptible in a new-born child.

" 3. The brain arrives, on an average, at its full size towards the seventh or eighth year. Soemmering says, erroneously, that the brain does not increase after the third year. Gall and Spurzheim, on the other hand, are of opinion that the brain continues to grow till the fourteenth year. The brothers Wenzel have shewn that the brain arrives at its full growth about the seventh year. This is confirmed by Hamilton's researches."

(The reader will perceive that these statements do not exactly accord with the results of Dr. John Reid's observations. It seems probable that the data upon which Tiedemann's conclusions were founded have been too limited in number. In calculating the weight of the brain in adolescence and adult age, some allowance should be made for the greater proportion of water at the former period; the quantities of that fluid being at those ages 72 and 74 parts in 100 respectively, according to L'Héritié.)

" 4. Desmoulins is of opinion that the brain decreases in old people. From this circumstance he explains the diminution of the functions of the nervous system and intellectual powers. The truth of this assertion has not as yet been determined. The brothers Wenzel, and Hamilton deny it.

"It is remarkable that the brain of a man, eighty-two years old, was very small, and weighed but 3 lbs. 2 oz. 3 dr., and the brain of a woman, about eighty years old, weighed but 2 lbs. 9 oz. 1 dr. I have generally found the cavity of the skull smaller in old men than in middle-aged persons. It appears to me, therefore, probable that the brain really decreases in old age, only more remarkably in some persons than in others.

"5. There is undoubtedly a very close connection between the absolute size of the brain and the intellectual powers and functions of the mind. This is evident from the remarkable smallness of the brain in cases of congenital idiotismus, few much exceeding in weight the brain of a new-born child. Gall, Spurzheim, Haslam, Esquirol, and others have already observed this, which is also confirmed by my own researches. The brain of very talented men, on the other hand, is remarkable for its size.

"Anatomists differ very much as to the weight of the brain compared with the bulk and weight of the body; for the weight of the body varies so much, that it is impossible to determine accurately the proportion between it and the brain. The weight of an adult varies from 100 to 800 lbs., and changes both in health and when under the influence of disease, depending in great measure on nutrition. The weight of the brain, although different in adults, remains generally the same, unaltered by the increase or diminution of the body. Thin persons have, therefore, relative to the size of the body, a larger brain than stout people.

"From my researches I have drawn the following conclusions.

"1. The brain of a new-born child is, relatively to the size of the body, the largest; the proportion is 1 : 6.

"2. The human brain is smaller, in comparison to the body, the nearer man approaches to his full growth. In the second year the proportion of the brain to the body is as 1 : 14; in the third, 1 : 18; in the fifteenth, 1 : 24. In a full-grown man, between the age of twenty and seventy years, as 1 : 35 to 45. In lean persons the proportion is often as 1 : 22 to 27; in stout persons, as 1 : 50 to 100 and more."

(This estimate, as far as regards the early ages, differs from that of Dr. John Reid, probably owing to the difference in the number weighed.)

" 3. Although Aristotle has remarked that the female brain is absolutely smaller than the male, it is nevertheless not relatively smaller compared with the body; for the female body is, in general, lighter than that of the male. The female brain is for the most part even larger than the male, compared with the size of the body.

" The different degree of susceptibility and sensibility of the nervous system seems to depend on the relative size of the brain as compared with the body. (qu. ?) Children and young people are more susceptible, irritable, and sensitive than adults, and have a relatively larger brain. Thin persons are more susceptible than stout. In diseases which affect the nourishment of the body, the susceptibility increases as the patients grow thinner. The susceptibility and sensibility decreases, on the other hand, with persons recovering from a long illness, gradually as they regain their strength. The degree of sensibility in animals is also in proportion to the size of the brain. Mammalia and birds have a larger brain and are more susceptible than amphibious animals and fishes."*

Enough has been said to show, that in contrasting the brain of man and that of the lower animals, with reference to the

* Tiedemann on the Brain of the Negro compared with that of the European and the Orang Otang. Phil. Trans. 1836.

much agitated question of the connection of mental faculties and intellectual endowments with that organ, no *one* standard of comparison must be selected. We must look to absolute and relative size—we must compare the bulk of the several portions of the encephalon with each other—we must notice the size of the encephalic nerves in relation to the whole organ —and, above all, we must compare the intimate organization of brain one with the other. Unless all the features of the brains that are subjected to comparison be carefully taken into the account, erroneous conclusions will be obtained. For instance, the brain of the elephant is absolutely larger than that of man: the convolutions of the hemispheres are very highly developed, and exhibit a degree of complexity almost equal to that of the human brain. At first sight we might be led to infer a very close approximation to the human, and place the elephant very high up in the scale of cerebral developement. In comparing, however, the brain of this animal with that of the monkey, the following result is obtained. The encephalon of the elephant is above that of the monkey by the superior developement of the cerebral convolutions; it is equal to it, as regards the quadrageminal bodies, but from the general form of the brain, the length of its transverse diameter, the presence of olfactory eminences, the position of the cerebellum (uncovered by the posterior lobes), it must be placed on a level with that of the inferior Mammalia.*

Of the brain in different races of mankind.—When so much diversity is observable in the form of the cranium in different races of mankind, it seems reasonable to expect a corresponding variety in the shape and other characters of the encephalon. The external form of this latter organ will correspond with that of the cranium, and its size with the capacity

* Leuret, op. cit. p. 448.

of that cavity. But it is plain that as the capacity of the skull is no wise necessarily affected by its shape, so the absolute bulk of the brain need not vary, although its containing case may exhibit much variety of form. The great question for the physiologist to determine is, whether, in the various races of mankind, the brain exhibits any striking peculiarities, characteristic of one or more of them, or whether it presents no more variety of shape, size, weight, and structure than may be observed in different individuals of any one of those races.

It should be premised that actual observations of the brain of different races are few. In Europe, where hitherto anatomy has chiefly been studied, the means of instituting such inquiries on a large scale have been altogether wanting. But it may be confidently expected that the many well-educated men who now visit distant climes, accompanying our fleets and armies, will not let slip the opportunities which they possess, without contributing somewhat to the solution of so interesting a question.

Many years ago it was thought that the brain of the coloured races possessed a greater quantity of colouring matter than that of the white, and this opinion appears to have originated with J. F. Meckel, who asserted that the grey substance was of a darker hue than in the European brain, and also that the medullary substance was not so white, but yellowish grey or light-brown.* Walter, Camper, Bonn, Soemmering, have however, amply refuted this statement.

Walter denied more particularly that part of the assertion which attributed a darker colour to the white substance. He states that it is just as white as in the European, but that the

* De la diversité de couleur dans la substance medullaire de Negres, Hist. de l'Acad. de Berlin, 1753. Du Cerveau des Negres, ibid. 1757, quoted in Tiedemann's paper.

cortical substance is darker, that is, of a greyish brown colour, which he attributed to the darker colour of the blood in the Negro.*

Soemmering, with a view to decide the question, dissected three perfectly fresh Negro brains in the presence of other anatomists, Professors Weichmann, Schumlanski of Petersburgh, and Billman of Cassel, taking the very proper precaution to compare on the spot the fresh brain of an European. The result was that he could not discover either the cineritious or medullary substance to be in the least darker than in Europeans; he even thought that the colour was rather paler in the African than in the European brain.†

It is true that Caldani and Rudolphi appear to have considered the grey substance darker in the Negro than in the European, the former having examined the brains of two Negroes, and the latter that of a Mulatto. But little dependence is to be placed on statements founded upon such a limited number of observations, and moreover it is well known that the aspect of the grey substance varies in different individuals according to the quantity of blood which it may contain.

Tiedemann affirms that the brain of the Negro does not present any material difference from that of other nations. Judging by Camper's rule, founded upon the measurement of the facial angle, which is smaller in the Negro than the European, it had been supposed that the latter was smaller. The results of a few cases in which the Negro brain was weighed do not confirm this statement. The brain of a Negro boy according to Soemmering weighed 2 lbs. 10 oz. 3 dr. avoirdupois, or 3 lbs. 6 oz. 6 dr. troy. The brain of a tall

* Epistola Anatomica ad W. Hunterum de venis oculi. Berolin. 1778.

† Vom Korperlichen Unterschied des Negers, p. 18.

handsome Negro, about twenty years of age, weighed 2 lbs. 13 oz. 4 dr. avoidupois, or 3 lbs. 9 oz. 4 dr. troy weight. A Negro's brain, examined by Sir Astley Cooper, weighed 3 lbs. 1 oz. or 49 oz. and that of a young Negro, aged twenty-five, short and thin, examined by Tiedemann himself, weighed 2 lbs. 3 oz. 2 dr., having been a short time kept in alcohol.

Tiedemann has also contrasted the capacity of the Negro skull with those of men of the Caucasian, Mongolian, American, and Malayan races. This was done by first weighing the skull with or without the lower jaw-bone. Then the skull was weighed, having been filled with dry millet seed through the foramen magnum. Lastly, by deducting the weight of the empty skull from that of the filled one, the capacity of the cranial cavity was obtained.

In the Ethiopian race, the range of capacity was found to be, in male skulls from 54 oz. 2 dr. 33 gr. to 31 oz. 5 dr. 16 gr. troy, in thirty-eight observations, and in female skulls from 31 oz. 4 dr. to 24 oz. 7 dr. 39 gr. in three observations.

In the Caucasian race, the capacity of male skulls of European nations was found to range between 57 oz. 3 dr. 56 gr. to 32 oz. 6 dr., in seventy-seven observations, and that of male skulls of Asiatic nations from 41 oz. 5 dr. 6 gr. to 27 oz. 6 dr. 30 gr. (a Hindoo Brahmin's head), in twenty-four observations.

The male skulls of the Mongolian race exhibited a capacity from 49 oz. 1 dr. 22 gr. to 25 oz. 0 dr. 18 gr. (a native of Nootka Sound), in eighteen observations.

In the American race the capacity of the male skulls ranged between 59 oz. and 26 oz. 1 dr. 44 gr. (a Toway Indian), in twenty-four observations.

And in the Malayan race it ranged from 49 oz. 1 dr. 45 gr.

to 30 oz. 5 dr. in thirty-eight observations, and in five female skulls from 37 oz. 5 dr. to 19 oz. 2 dr. 49 gr. (a Lascar woman.)

These researches certainly give no countenance to the doctrine which assigns the lowest place, in the chain of human varieties, to the Negro as regards cerebral developement. So far is this from being the case, that the Ethiopian race differs to a very trifling degree from the European; and, indeed, the examples of skulls of the smallest capacity are found among Asiatic natives (Hindoos) and Americans.

The following conclusions are derived by Tiedemann from his comparison of the Negro brain with that of other races.

" 1. The brain of a Negro is upon the whole quite as large as that of the European and other human races.

" 2. The nerves of the Negro, relatively to the size of the brain, are not thicker than those of Europeans, as Soemmering and his followers have said.

" 3. The outward form of the spinal cord, medulla oblongata, the cerebellum and cerebrum of the Negro shows no important difference from that of the European.

" 4. The Negro brain does not resemble that of the orang otang more than the European brain does, except in the more symmetrical distribution of the gyri and sulci. It is not even certain that this is always the case. We cannot therefore coincide with the opinion of many naturalists, who say that the Negro has more resemblance to apes than Europeans in reference to the brain and nervous system. It is true that many ugly and degenerate Negro tribes *on the coast* show some similarity in their outward form and inward structure to the ape; for instance, in the greater size of the bones of the face, the projecting alveoli and teeth, the prominent cheek-bones, the recession of the chin, the flat form of the nose-bones, the projecting and strong lower jaw, the position of the foramen occipitale magnum, the relative greater length of the ossa humeri and the bones of the foramen, the flat foot, and in the length, breadth, shape, and position of the os calcis. * * * These points certainly distinguish many Negro tribes from the Europeans, but they are not common to all the Negroes of the interior of Africa, the greater number of which are well made, and have handsome features."*

A series of researches so extensive and conducted with so much care, (although the actual comparison of the brains themselves is yet wanting,) cannot allow a doubt to arise as to the conclusion which ought properly to flow from them. It would appear from them that no very marked differences exist between the brains of any of the classes of mankind—that the same relative inferiority of women to men is universally met with—and that a very diminutive state of brain may be, when not an accompaniment of idiotcy, either a part of a frame originally very small in stature, or a degenerate condition consequent upon a life of the lowest barbarism, under every possible physical impediment to the developement of bodily vigour, wholly deprived of moral or intellectual culture, a state which becomes more and more degenerate in each succeeding generation, or, lastly, the effect of the mechanical compression to which many tribes subject the crania of their offspring in early infancy.

In proceeding to the examination of the human encephalon, it seems expedient to premise a few observations on the method which it is most advisable to adopt for this purpose.

* The remaining observations of Tiedemann on the intellectual condition of the Negro merit attentive perusal. See also Prichard on the Physical History of Mankind, vol. i. p. 197, and vol. ii. p. 346.

The inferior limit of the encephalon is the plane of the occipital foramen. In examining that great nervous mass which is situate above this plane, it will be obvious, even to the most superficial observer, that it admits of a convenient subdivision into certain great segments, each of which, although

Fig. 20.

Section in the vertical direction, to show the relation and mode of connection of the various segments of the encephalon. (After Mayo.)

g, fibres passing to the posterior lobe of the brain; g, corpus geniculatum externum; n, anterior of the corpora quadragemina (nates); b, posterior of corpora quadragemina (testes); f, olivary fascicles; o, olivary bodies; v, pons Varolii; p, anterior pyramids; r, restiform bodies (forming part of the crus cerebelli); t, processus cerebelli ad testes; c, cerebellum; s, spinal cord.

extensively connected with the neighbouring ones, may yet be capable of acting as an independent centre, and, in short, possesses the anatomical as well as physiological properties of a ganglion. And on a more minute investigation the number of gangliform segments will be found to be greater than the observation of the mere surface of the encephalon would lead us to conclude. The subdivision, however, which it is most convenient for the purpose of description to adopt, is that already stated at page 77, into, 1, *the medulla oblongata*, which is immediately continuous inferiorly with the spinal cord. This segment has certain characters of structure which decidedly indicate its ganglionic nature; several nerves of considerable size and of great physiological importance are implanted in it, and its external anatomy very clearly indicates its distinctness from the spinal cord inferiorly and from the other encephalic segments above, of which that next in order proceeding from below upwards, is, 2, the *mesocephale*. To this mass, so called because of its intermediate position between the other segments, the term *isthmus* has been also very appropriately applied, as it is the connecting link between all the encephalic segments. Inferior and posterior to it is placed, 3, the *cerebellum*, which has very intimate relations to the medulla oblongata as well as to the segment last described, but much less extensive ones to that which forms by far the largest proportion of the encephalon namely, 4, the *cerebrum*, which therefore occupies the principal portion of the cranial cavity.

The distinction between these different segments is very obvious on an examination of the surfaces of the brain, which indeed ought to be the first step to be taken by the anatomist. To discover how they are connected to each other and to the spinal cord, how the corresponding portions on opposite sides of the mesial plane are associated together, what fibres are common to all the segments, and what peculiar to some, and, lastly, how the grey matter is related to the white,—these are the chief objects to be attained in the dissection of the brain. No one method of dissection will suffice for this purpose. The anatomist should first make himself familiar with the simple topographical anatomy of the brain, that is, with all those parts in it which possess such characters of form or structure as may entitle them to be regarded as distinct and deserving of separate description, and have obtained for them a special appellation. The form, size, general structure, and relations of these parts should be carefully noted. And this method of examination is equally applicable to the dissection of each segment of the encephalon. But the most convenient way in which it can be conducted for ordinary practical purposes, is to commence with the cerebral hemispheres, and having studied their general structure as displayed on a horizontal section, to examine the extent and connections of the fibres which connect the right and left hemispheres with each other (the corpus callosum); then to open the ventricles, examine their shape and extent, and note the various particulars connected with the numerous parts which are brought into view by exposing those cavities. The dissector may next observe how certain of the parts concealed by the lateral ventricles are connected with the mesocephale (the optic thalami for instance), and, having been already acquainted with the various prominences which are seen upon the surface of the latter, he may by vertical, or transverse, or horizontal sections, investigate the manner in which the white matter of this segment connects itself with that of the neighbouring ones. In examining the cerebellum, the larger fissures afford sufficient indication for a convenient subdivision of the organ, and by horizontal or vertical sections at various parts of it the connection of the grey and white matter may be displayed, and of the latter to the mesocephale and medulla oblongata. The medulla oblongata has upon its surface various lines or fissures which denote the proper limits of its constituent columns, and which will be sufficient guide to the dissector in tracing the extent and connections of each. Transverse and longitudinal sections also afford useful information respecting the structure of this segment of the encephalon and the relations of its parts.

Such is the mode of dissection from above downwards, against which it has been greatly the fashion of late years to declaim with much vehemence. But, however the advocates of a particular theory may object, there can be no doubt that this method is by far the most useful for all practical purposes. It enables the anatomist, without difficulty, to study the prominent parts or landmarks (so to speak) of the brain, without a knowledge of which it is in vain to attempt any other mode of dissection. And for pathological investigations it is the only method which can be conveniently adopted. It is plain, therefore, that all who are desirous of becoming acquainted with the anatomy of this organ should begin by making dissections in this way. An additional advantage is found in this mode of investigation, from its great applicability to the dissection of the brains of the lower animals, of the Mammalia and Birds especially, for the purpose of comparing them with the brain of the human subject.

The method of our celebrated countryman Willis was very much the same as that above described. He removed the membranes from the posterior lobes of the hemispheres, and thus separated the latter from the subjacent parts, and by raising them as far forwards as possible he was enabled to observe the connections of the cerebral hemispheres with the mesocephale, and the attachments of the fornix behind. He also must have studied the substance of the hemispheres by horizontal section. By then dividing the posterior parts of the hemispheres horizontally along the plane of the corpora striata, he raised a large flap consisting of the upper part of the hemispheres, with the intervening corpus callosum and the adherent fornix; and thus were exposed the inferior surface of the latter, and the cavities of the three ventricles, the fourth being shewn by a vertical section of the cerebellum on the median plane, and by the separation of the segments thus made. This is an admirable section to display the connection of the hemispheres with what Willis described as the medulla oblongata, namely, in the words of his translator, " all that substance which reaches from the inmost cavity of the callous body and conjuncture in the basis of the head to the hole of the hinder part of the head, where the same substance being yet further continued ends in the spinal marrow." The fourth, seventh, and eighth plates in Willis's work display this mode of dissection.*

The modern researches of Reil, Gall and Spurzheim, and others, directed attention more particularly to the physiological anatomy of the brain. Their principal object was to discover the mode of connection of the several segments of the cord with each other, and of the whole encephalon with the spinal cord. And their method of dissection consisted in tracing the course of the fibres chiefly from below upwards. Reil found it necessary to harden the brain in alcohol, in order to give it such firmness as would enable him to tear portions of it in the direction of its fibres, and thus to make these latter conspicuous. There can be no doubt that layers of the brain will

* Thomæ Willis, Cerebri anatome, nervorumque descriptio et usus, in Opera Omnia, Amsterdam, 1682, cap. xiii. Also an English edition by S. Pordage, London, 1684. The following extract gives the description of Willis's dissection in his own words. " Ut cerebri ita proprie dicti anatome rite celebretur, haud vulgari sectionis modo procedendum esse existimo. Verum ubi totius εγκεφαλου calvaria exempti compages coram sistitur, imprimis posterior cerebri limbus, ubi cerebello ac medullæ oblongatæ con-

separate most readily when torn in the direction of their fibres; and thus this mode of preparation becomes of great importance to the anatomist, as he can thereby determine easily the direction of those fibres which form the principal portion of the part under examination. It will not, however, suffice to display the direction of all the fibres, nor indeed is any mode of preparation adequate for that purpose; which can only be accomplished by extensive and patient microscopic investigation.*

nectitur, membranis undique discissis aut avulsis, a cohæsione cum partibus subjectis (quantum fieri potest) liberetur; tunc facile constabit quod cerebri substantia corporibus istis haudquaquam unitur, verum per se, nisi quod membranarum nexu superficie tenus conjungitur ab iis omnino libera ac independens fuerit: quinetiam hæc cerebri puppis a vicinis partibus eo ritu divisa, si anterius reclinetur, medullæ oblongatæ crura, prorsus nuda, ac a cerebro et cerebello (nisi in quibus locis hæc illi appenduntur) omnino distincta apparebunt. * * * * * * * * * * * * * Interioris cerebri recessus adhuc clarius patebunt, si limbus ejus a medullæ oblongatæ cohæsione, quantum fieri potest, ex omni parte separatus et elevatus, ad latera ejusdem medullæ, quibus juxta corpora striata unitur, paulo ulterius per substantiam suam secetur, simulque fornix juxta radices discissus una cum cerebro reflectatur, tunc enim cerebri compages penitus elevari, antrorsum reflecti, ac in planum explicari potest, ita ut corporis callosi in aream latam expansi interior superficies tota conspici et tractari possit. Ubi, præter medullarem et nitidissimam illius substantiam, observare est plures lineas albas paralelas quæ cerebri dissepimentum rectis angulis secant; quasi essent tractus quidam, sui vestigia, in quibus spiritus animales ab uno cerebri hæmispherio in alterum migrant resiliuntque." Op. cit. cap. i. p 5, 6.

* Reil's methods of preparing the brain are best described in his own words: "Of the methods which I have employed in preparing brains, those contained in the following directions answered best. 1. Let the brain be hardened in alcohol, and then placed in a solu-

The great advantage of pursuing the dissection in the direction from below upwards consists in this, that we proceed from the more simple to the more complex. The problem which the anatomist has to solve is, Given certain columns or bundles of fibres in the medulla oblongata, to determine how they connect themselves with the other segments of the brain. But it is obvious that without some knowledge of the topography of the other more complicated parts of the encephalon, the dissector would have considerable difficulty in pursuing his researches. Nor must he content himself with the solution of this fundamental question; he is to explore for other fibres in these segments besides those which connect them with the medulla oblongata, and he has to ascertain how they comport themselves, whether as forming an integrant portion of the segment in which they are found, or serving to connect it with one or more of the others.

tion either of carbonated or pure alkali, in the latter two days, in the former for a longer period, and then again hardened in alcohol if thus rendered too soft. The advantage of this method is, that the fasciculi of nervous matter are more readily separable, and the brown matter more distinguishable from the white than after simple maceration in alcohol; the grey matter is rendered by the alkali of a blacker grey, and assumes the consistence of jelly. 2. Let the brain be macerated in alcohol, in which pure or carbonated potass or ammonia has been previously dissolved; the contraction of the brain is lessened by this process. 3. Let the brain be macerated in alcohol from six to eight days, and then its superficial dissection commenced, and the separation of the deeper parts continued, as the fluid, in which the brain is kept immersed, penetrates its substance. This method appears to me better than the preceding, and would very likely be improved if the alcohol were rendered alkaline. The fibres in a brain, thus prepared, are more tenacious than otherwise, and the deeper parts are sooner exposed to the influence of the alcohol."—Mayo's translation of Reil's Eighth Essay, in the former's Anat. and Phys. Commentaries, p. ii. p. 50.

Although we are mainly indebted to modern anatomists for following out more completely this method of dissection, it cannot be denied that such men as Willis, Vieussens, and Malpighi were quite alive to the importance of examining the fibres of the brain, with a view to the physiological action of its different parts. No one can peruse Willis's admirable account of the brain without perceiving how completely he unites structure and function, and with what ingenuity he ascribes the passage of the nervous force (under the name of animal spirits) from one part to another, to the anatomical relations of those parts, and the direction of the constituent fibres. And, indeed, we may find in the writings of this great man the germs of many a theory which, in our own times, has been brought forward with a more plausible aspect, disencumbered of the quaint phraseology and superabundant metaphor so common in his day. I shall quote one remarkable instance as very much in point. Speaking of the fourth pair of nerves as connected with the corpora quadragemina, he says: " Concerning these little nerves it is observed, that when (although) many others proceed from the sides or the basis of the oblong marrow, these arise from the aforesaid Prominences in the bunching forth at the top (nates and testes). The reason of which, if I be not mistaken, is this,—we have affirmed that these prominences do receive and communicate to the brain the natural instinct delivered from the heart and bowels to the cerebel; and on the other side, or back again, do transfer towards the Præcordia, by the mediation of the cerebel, the forces of the passions or affections received from the brain; but in either action the motion of the eyes is affected with a certain manifest sympathy. For if pain, want, or any other signal trouble afflicts the viscera or the præcordia, a dejected and cast-down aspect of the eyes will declare the sense of its trouble: when on the contrary, in joy, or any pleasant affection of the præcordia or viscera, the eyes are made lively and sparkle again. In like manner the eyes do so clearly show the affections of the mind, as sadness, anger, hatred, love, and other perturbations, that those who are affected, though they should dissemble, cannot hide the feeling and intimate conceptions of the mind. Without doubt these so happen because the animal spirits tending this way and that way in this deviating place between the brain and the præcordia, do at once strike those nerves as the strings of a harp. Wherefore, from this kind of conjecture, which we have made concerning the use of these nerves, we have called them *Pathetical*, although indeed other nerves may deserve the same name."*

Malpighi and Vieussens were well acquainted with the fibrous structure of the brain, and appear to have had very correct notions as to the general direction which they assume, and the parts which they serve to connect to each other. The former describes the fibres of the brain and cerebellum as taking their origin from the top of the spinal marrow contained within the cranium (*medulla oblongata*); " for they ramify from four reflected crura of this medulla in all directions, until they end by their branched extremities in the cortex." Vieussens states that the medullary substance is composed of innumerable fibrils connected together and arranged into various fasciculi, which become very obvious when it is boiled in oil.†
The great merit of Reil, Gall and Spurzheim, and their followers in later years, consists in their having followed out with great diligence the coarser anatomy of those fibres, and determined many important and undeniable truths. But in the

* English edition of Willis's Works, p. 90, fol. Lond. 1684.

† Malpighi, Exercitatio Epistolica de Cerebro, 1664. Vieussens, Neurographia Universalis, lib. i. cap. x. See the whole passages quoted in Dr. Gordon's Observations on the Structure of the Brain. Edin. 1817, p. 21.

statements of all anatomists, who avail themselves of no other aid than that which the naked eye affords, there is much that must necessarily be uncertain or doubtful, nor is there any other mode of removing these uncertainties but by the successful application of microscopical analysis to the whole cerebral structure.

Of the surface of the encephalon.—We now proceed to examine the various points worthy of notice in the superficial anatomy of the encephalon.

The shape of the brain is determined by that of the cerebral hemispheres. A line drawn around the surface of the latter, so as to enclose them, would describe an oval, the smaller extremity of which is directed forwards.

The superior and lateral surfaces of the encephalon are convex, and have a smooth appearance from the visceral layer of the arachnoid being extended over them, adhering to the subjacent pia mater. When the membranes have been removed, the convoluted character of these surfaces, previously seen through them, becomes very manifest, as will be more particularly described by-and-bye. The longitudinal and transverse diameters of these surfaces correspond to those of the cranial cavity.

The superior surface is divided along the median plane into two equal and in a great degree symmetrical portions by a fissure which passes vertically between them and receives the great falciform process of the dura mater. In front and behind, this fissure completely divides the cerebral lobes. In the latter situation, the tentorium cerebelli is seen at the bottom of it when the hemispheres are separated, if the encephalon be *in situ;* if it have been removed, however, the superior surface of the cerebellum forms the floor of the fissure. In the middle the fissure is interrupted by a horizontal lamina of white fibres, which is called the *corpus callosum*, the great commissure of the cerebral hemispheres.

Fig. 21.

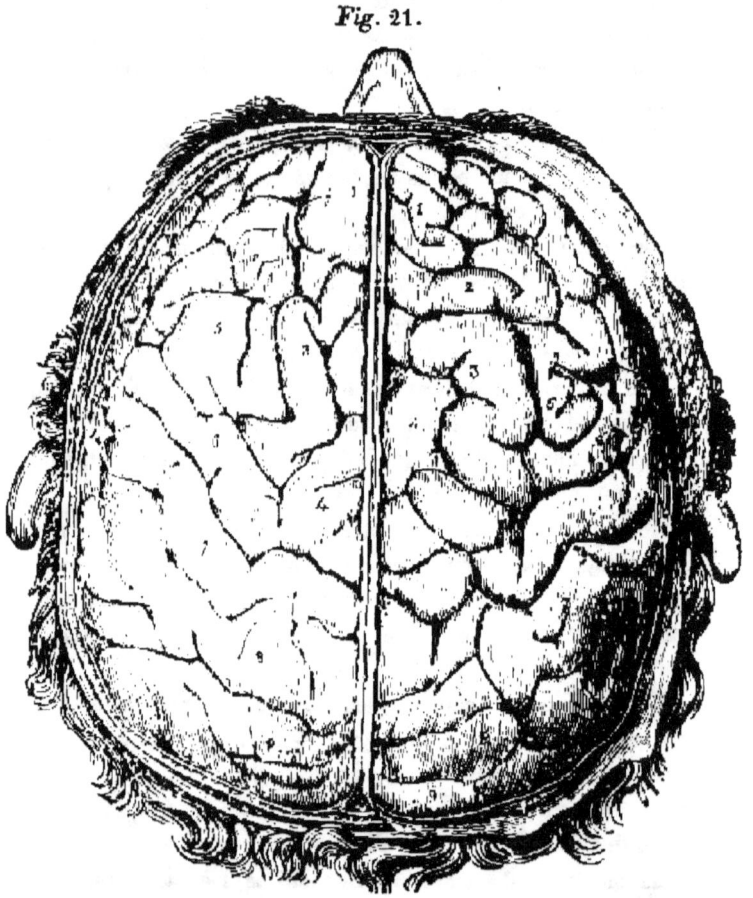

The superior and part of the lateral surfaces of the encephalon, exposed by the removal of the calvaria. The falx cerebri is seen in the longitudinal fissure. The figures on the convolutions indicate those of opposite sides which present some degree of symmetrical character. They will be referred to further on in the description of the hemispheres of the brain.

The inferior surface of the encephalon, called commonly the *base of the brain*, also presents many points worthy the attention of the anatomist.

It is not all upon one level: in this respect it corresponds with the disposition of the base of the skull. We find, indeed, three segments, each on a different plane, and corresponding to each of the three fossæ of the cranium. This is best observed by examining a vertical section of the head, the brain being retained in its situation, or by removing the wall of the cranium on one side quite down to its inferior surface.

The anterior segment, and that which is on the highest level, corresponds to the anterior fossa of the cranium. It rests, therefore, upon the roofs of the orbits, and its surface is on each side slightly concave to adapt it to the form of its resting-place. The continuation of the anterior median fissure separates its right and left portion, and the attachment of the falx to the crista galli of the ethmoid makes the distinction more complete. In a distinct sulcus, parallel to and immediately on each side of the longitudinal fissure, we find the olfactory process or nerve. This segment forms the inferior surface of what anatomists commonly designate as the anterior lobes of the brain. It presents the convoluted appearance which is conspicuous on the proper cerebral surface every where. A curved fissure of considerable depth, called the *fissure of Sylvius*, is the posterior limit of each anterior lobe.

The fissure of Sylvius corresponds on each side to the posterior concave edge of the lesser ala of the sphenoid bone, which is received into it. It may be traced from within, commencing at a triangular flat surface (*locus perforatus anticus*), which corresponds to the posteror extremity of each olfactory process. From this situation it proceeds outwards and curves backwards and a little upwards; its convexity is therefore directed forwards. Towards the lateral surface of the brain it becomes continuous with the fissures of neighbouring convolutions.

The fissure of Sylvius is of considerable depth, especially at its internal extremity, and, like all the fissures of the brain, large or small, is lined by the pia mater. We notice here a large interval between the arachnoid and pia mater, in which a considerable accumulation of the cerebro-spinal fluid takes place, communicating with the anterior conflux of that fluid. In this space runs the middle artery of the brain, giving off its branches to the sides and floor of the fissure. When the convolutions which bound the fissure are separated, a variable number of small convolutions is found, projected from its floor as an insulated lobe, which is enclosed by a bifurcation of the fissure. This lobe constitutes the island (*insel*) of Reil.

The *middle* segment which lies immediately behind the Sylvian fissure, is on a plane much lower than the anterior, and corresponds on either side to the deep and hollow median fossa of the cranium. It consists of two lateral very convex lobes, commonly known as the *middle lobes of the brain*, which are separated from each other by a deep depression. These lobes, which are very accurately limited in front by the fissure, have no exact boundary behind, but pass off very gradually into the *posterior lobes* of the hemispheres, as may be seen by raising up the cerebellum.

The transition from the middle to the posterior lobe of the hemisphere is only indicated by the different character of the inferior surface of the hemisphere, the former being convex, the latter concave. The subdivision, indeed, of the cerebral hemisphere into middle and posterior lobes is purely conventional, and I agree with Cruveilhier that it ought to be discarded, for it has no foundation in the anatomy

BASE OF THE BRAIN. 141

Fig. 22.

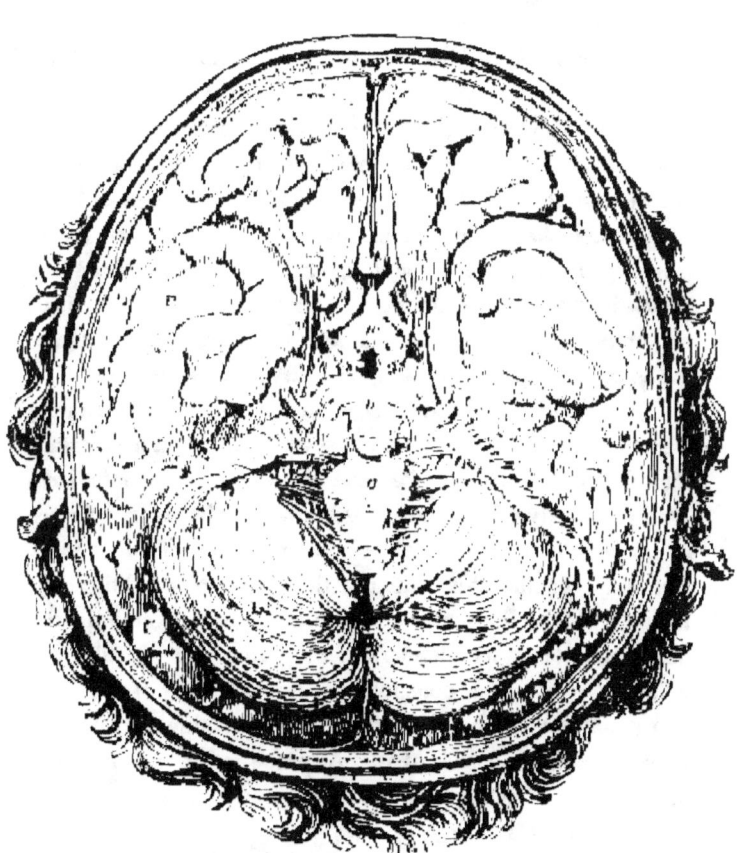

Base of encephalon viewed from below.

A, anterior lobe ; B, middle lobe; C, posterior lobe; D, cerebellum ; *a,* olfactory processes ; *b,* optic nerves; *c,* third pair of nerves; *d,* fourth pair of ditto ; *e,* fifth pair—portio major; *e',* fifth pair—portio minor ; *f,* sixth pair ; *g,* seventh pair ; *h,* filaments of origin of the glosso-pharyngeal and vagus ; *i,* spinal accessory nerve ; *k,* ninth nerve ; *l,* pituitary body and process proceeding from the tuber cinereum ; *m,* mamillary bodies; *n,* pons Varolii ; *o,* medulla oblongata.

of the parts. The whole of that portion of the cerebral hemisphere which is situate behind the Sylvian fissure should be called the posterior lobe.

The hollow space between the middle lobes of the brain corresponds to the principal anterior reservoir of subarachnoid fluid. It is situate immediately above the Sella Turcica, and, indeed, the brain is, as it were, tied to the pituitary body, which is firmly lodged in this excavation of the sphenoid bone, by a funnel-shaped hollow process of nervous matter, called *pituitary process* or *tube,* (*m, l, fig.* 22), which, enveloped in a sheath of arachnoid membrane, is inserted into it by its small extremity. This space communicates with the anterior fissure in the middle, and with the Sylvian fissure on either side.

Commencing at the anterior fissure and passing backwards, we notice the following parts, to see which clearly it is necessary that the adherent pia mater and the arachnoid should have been previously carefully dissected away. The anterior fissure is limited by the anterior fold or reflection of the corpus callosum: behind this we find a thin layer of a lightish grey matter, which, like a triangular plate, seems to stop up the third ventricle at its inferior surface. This, indeed, which is called *tuber cinereum,* constitutes a principal part of the floor of that ventricle. The pituitary process is continuous with and is probably an extension of it. A probe introduced into the cut extremity of this process will be found to pass readily into the third ventricle.

Immediately in front of the pituitary process, the union of two white bands, which form lateral boundaries to a large portion of the tuber cinereum, *the optic tracts,* takes place along the middle line. This forms the *commissure* of the optic nerves, from which these nerves diverge. Behind the pituitary process the tuber cinereum extends back to two small pisiform bodies of an extremely white colour on their surface, *corpora mamillaria* or *albicantia* (*m*, *fig.* 22). These, we shall see by-and-bye, are connected with one of the most important of the cerebral commissures, namely, the *fornix*.

Behind the mamillary bodies we find a deep depression into which the pia mater sinks, carrying with it very numerous bloodvessels. This depression lies between two thick processes of fibrous matter, which, traced from below, pass upwards and outwards, expanding as they advance, and upon which each hemisphere is placed (to use Reil's simile) like a mushroom on its stalk. These are the *crura cerebri*, the peduncles of the cerebral hemispheres. The depression above described, which separates them, is the *intercrural* or *interpeduncular* space. When the pia mater has been removed from it, its surface appears cribriform from the perforations of the numerous minute vessels which penetrate it; it has been named by Vicq d'Azyr *substantia perforata media*. The nervous matter which forms the floor of this space has a greyish hue, and connects the crura to each other, like a bridge, whence the designation *pons Tarini*. At the interpeduncular space we see the third pair of nerves emerging from their connection with the crura cerebri.

The inner margin of each middle lobe of the brain is separated from the corresponding crus cerebri by a fissure which passes from behind forwards, and terminates in the fissure of Sylvius. If this fissure be followed backwards, it will be found to become continuous with a transverse fissure which separates the cerebrum from the cerebellum, and corresponds to the posterior edge of the corpus callosum. A continuity is thus established between the lateral and the transverse fissures, whence results one great fissure of semicircular form, the concavity of which is directed forwards. This is the *great cerebral fissure* of Bichat, or the great *transverse* or *horizontal fissure* (Cruveilhier.) It may be described as commencing at the fissure of Sylvius on one side, turning round the opposite cerebral peduncle, and ending at the opposite Sylvian fissure. The anterior and lateral portions of this fissure have already been noticed as the situations at which the pia mater enters the brain to form the choroid plexuses of the lateral ventricles. And it may be remarked here, how freely the subarachnoid fluid may pass along this fissure from before backwards. Parallel to this fissure we find the fourth pair of nerves as it passes to its point of exit from the cranium.

Not the least interesting and important of the objects presented at this central portion of the base of the brain is that remarkable arterial anastomosis, called the *circle of Willis*. This will be more particularly described by-and-bye; but it may be stated here, that the anterior bifurcation of the basilar artery is immediately behind the interpeduncular space, on each side of which the posterior cerebral artery passes for a short distance. The posterior communicating artery is parallel to the inner edge of the middle lobe; the subdivision of the carotid corresponds to the commencement of the Sylvian fissure; and the anterior communicating artery is at right angles with the longitudinal fissure immediately behind the anterior reflection of the corpus callosum. This anastomosis of arteries is bathed in the liquid which occupies the subarachnoid space in this situation.

The tentorium cerebelli is situate on a plane a little beneath that of the middle segment of the base of the encephalon just described. It forms a septum between the posterior lobes of the cerebral hemispheres, which are continuous with the middle segment, and the posterior segment of the encephalon, which we now proceed to describe.

The posterior segment, as occupying the posterior fossa of the cranium, is on a level considerably below that of the middle segment. The parts which are deserving of more particular notice here, are, proceeding from before, the pons

Varolii (*n, fig.* 22), the inferior and anterior surface of the mesocephale, which is situate immediately behind the interpeduncular space, the crura cerebri appearing to emerge just above its anterior border. From its posterior edge the medulla oblongata (*o*) extends downwards, and a little backwards. As the brain rests on the upper surface of its hemispheres with its base upwards, the medulla oblongata is seen to occupy a notch or depression between the hemispheres of the cerebellum. The fibres of the pons Varolii are seen passing outwards and backwards into each hemisphere of the cerebellum, forming the inferior layer of each crus cerebelli. On each side of the medulla oblongata is the inferior convex surface of each hemisphere of the cerebellum marked by its fissures and laminæ. The basilar artery passes in a groove along the middle of the pons from before backwards. The fifth nerve emerges from the crus cerebelli, the sixth immediately below the posterior margin of the pons, and the seventh, eighth, and ninth nerves are seen springing from each side of the medulla by a series of fascicles similar to those which form the roots of the spinal nerves.

Of the dissection of the brain from above downwards.— It will facilitate our subsequent descriptions, if, previous to examining the several segments of the encephalon in detail, I give a rapid sketch of the dissection of the brain according to the *topographical* method, proceeding from above downwards.

This dissection is commenced by making a horizontal section of one hemisphere, a little above the level of the corpus callosum. The surface, which is thus exposed, has in shape the character of a demi-oval. It is chiefly composed of white substance, which occupies the centre of the space, bounded by a wavy border of grey matter. Anatomists designate it *centrum ovale minus*. We find this a convenient section on which to study the anatomy of the convolutions, and to give some idea of the composition of that portion of each hemisphere of the brain which is situate above the ventricles. On making a similar section of the other hemisphere at the same level, a similar surface is exposed, and the conjunction of both constitutes what Vieussens denominated the *centrum ovale majus*.

By separating the hemispheres slightly, after this section, the horizontal portion of the corpus callosum is well displayed. The continuity of its transverse fibres with the white substance of the hemispheres may be traced; and by following its anterior and posterior reflections they will be found to connect the hemispheres at their inferior as well as their superior parts. The corpus callosum, when examined in its full extent, exhibits somewhat of a vaulted shape, and is found to enter largely into the formation of the roof of the lateral ventricles.

We notice some remarkable longitudinal fibres, passing along the middle of the corpus callosum, varying greatly in developement in different brains. These consist of two bundles placed in juxta-position, but easily separable. We may trace them throughout the whole length of the corpus callosum. They cut the transverse fibres at right angles, and may be readily dissected up from them. They seem to tie the transverse fibres together, and are probably commissural. They form what has been improperly called *the raphé* of the corpus callosum, more correctly *the longitudinal tracts* (Vicq d'Azyr).

By scraping away the white substance, on each side of the corpus callosum, the lateral ventricles may be opened. If this be done with great care, a considerable portion of the membrane which lines the interior of each ventricle may be exposed, but such is its great delicacy that a very slight force ruptures it. When there is fluid in the ventricles, this membrane may be more easily demonstrated from its floating upon the fluid. The place at which the ventricles may be most certainly opened without the risk of injuring any of the parts contained within them, is about the eighth of an inch external to the apparent blending of the fibres of the corpus callosum with the white substance of the centrum ovale. With the handle of a knife the fibrous matter which forms the roof of the ventricle may be torn through in the antero-posterior direction, and the cavity thereby exposed.

Each lateral ventricle consists of a horizontal and a descending portion. The former resembles in shape an inverted italic S. Its *anterior* extremity, or *cornu*, is directed outwards; the *posterior* turns inwards towards that of the opposite side. The *descending cornu* passes downwards, forwards, and inwards in a curved course with the concavity forwards and inwards, and terminates at the fissure of Sylvius. The first has been appropriately designated the *frontal* ventricle, the second the *occipital*, and the third the *sphenoidal*, from their relations to the bones after which they have been respectively named. The posterior cornu is also named the *digital*, or *ancyroid* cavity.

The anterior cornua of the lateral ventricles are separated from each other by a vertical septum situated on the median plane, very thin and transparent, the *septum lucidum*. This may be easily demonstrated on a vertical section of the brain, made a little to one side of the mesial plane, or if both lateral ventricles have been opened, by supporting the corpus callosum on each side with the handle of a knife, by which means the septum is stretched, and its extent and connections may be more readily determined. The septum is of a triangular form with curvilinear base, which is directed forwards, and fits into the anterior reflection of the corpus callosum. Posteriorly it fits in between the corpus callosum and the anterior extremity of the horizontal portion of the fornix.

The septum lucidum, although so extremely delicate and transparent, is very obviously composed of two layers, enclosing a space or cavity called the *fifth ventricle*, which may be shewn by dividing the septum horizontally from behind forwards. Each of these laminæ consists of four layers, as may be easily observed by examining the margin of the section: the outer one is derived from the lining membrane of the ventricles; immediately within this is a layer of a pale greyish matter continuous with a similar layer which covers the optic thalami and the internal surface of the third ventricle, consisting of clear nucleus-like particles homogeneous in texture; a third layer is composed of white or fibrous matter; and a fourth consists of an extremely delicate membrane, probably covered by ciliated epithelium, which lines the internal surface of the fifth ventricle.

The fifth ventricle is closed at every point, and has, therefore, no communication with the lateral or other ventricles. It has been regarded by some as resulting merely from the artificial separation of the laminæ of the septum lucidum. And it seems unlikely that in life, during health, the surface of these laminæ should be otherwise than in contact, lubricated, however, by a slight moisture exhaled by the membrane. In a few rare cases fluid has been known to accumulate in this cavity.

In the fifth month of uterine life, according to Tiedemann, this ventricle communicates with the third through a small space, situate between the anterior pillars of the fornix and above the anterior commissure, and indeed it may be looked upon as a portion of the latter ventricle closed off by the formation of the fornix and septum lucidum.

The following parts are to be noticed in each lateral ventricle:—1. In the anterior horn, the *corpus striatum*, a pear-shaped eminence, the obtuse extremity of which is directed forwards and inwards. Posteriorly this body is apparently prolonged backwards into the inferior cornu of the lateral ventricle by a long tapering process which terminates there. 2. Internal and posterior to the corpus striatum is the *optic thalamus*, a gangliform body of a greyish colour, but considerably paler than that last named. 3. These two bodies are separated from each other by a superficial groove, in which lies a delicate band of fibrous matter, the *tænia semicircularis*, which is covered by a lamina of horny-looking matter, *lamina cornea*, the formation of which is attributed by some to a thickening of the lining membrane of the ventricle along this groove.

The choroid plexus in a great degree covers and conceals from view the optic thalamus. It passes up from the descending cornu, and just behind the septum lucidum and anterior pillars of the fornix turns inwards to unite with its fellow of the opposite side. On its inner side it is slightly overlapped by the thin margin of the horizontal portion of the fornix.

In the posterior horn we observe, on its internal wall, a projection inwards of one of the convolutions to which the name *hippocampus minor*, or *ergot*, has been given. It is an internal convolution, covered by a layer of fibrous matter derived from the fornix. It is traversed by a deep sulcus which may be exposed by cutting it across.

The descending horn contains a remarkable prominence, the *hippocampus major*, (also called *cornu Ammonis*,) which projects into it from its inferior wall, and follows the curve of the horn. It likewise may be regarded as an internal convolution, and is covered by a layer of fibrous matter derived from the fornix, which overlaps the concavity of the hippocampus by a thin margin, called *corpus fimbriatum*. Beneath this is a peculiar disposition of grey matter connected with the hippocampus, to which the name *fascia dentata* has been given. The commencement of the choroid plexus is found in this horn.

The anterior extremity of the descending horn of the lateral ventricle corresponds with the posterior extremity of the fissure of Sylvius. It is closed, not by nervous matter, but simply by the reflection of the membrane of the ventricle on the choroid plexus. This is the only provision against the escape of fluid from the ventricle. It seems highly probable, as we have already intimated, that there may be a communication at this situation, as well as at the fourth ventricle, between the fluid of the ventricles and that of the sub-arachnoid cavity by endosmose and exosmose. And the delicacy of the barrier which is opposed to the escape of fluid from the ventricle explains the occurrence of sanguineous effusions at the base of the brain from the rupture of vessels within the ventricle.

Postponing the more minute description of the parts found in the lateral ventricle, as above enumerated, we proceed with the examination of those which are brought into view beneath the corpus callosum.

The corpus callosum, which we have seen to consist of bundles of transverse fibres, passes directly from one hemisphere to the other. At its anterior and posterior extremity it is folded downwards, so as to connect those parts of the hemispheres which lie on a plane inferior to the lateral ventricles. Its anterior reflected portion, therefore, contributes to form the floor of the anterior horn, and the posterior one mingles with the fibres of the inner wall of the posterior horn. This disposition of the corpus callosum is best seen on a vertical section of the brain, which shows the vaulted form of this body. The greater abruptness of reflection of its posterior than of its anterior extremity, however, impairs in a great degree this character.

Of the fornix.—We have seen that the anterior reflection of the corpus callosum is occupied along the median plane by the vertical *septum lucidum*. This septum rests posteriorly upon the apex of a horizontal stratum of fibrous matter which forms part of a series of fibres called the *Fornix* or *Vault*. It is inconvenient to change names which have long been in use, more especially when there is no very certain scientific foundation for the adoption of a new one; otherwise the term *antero-posterior commissure*, which is suggested by the direction and the extensive connection of its fibres, might be appropriately assigned to it.

The principal portion or body of the fornix lies immediately beneath the three posterior fourths of the corpus callosum. By cutting this latter body across just at the posterior extremity of the septum lucidum, and dissecting the anterior segment forwards, and the posterior one backwards, its horizontal portion is exposed. In this dissection it is found that the latter portion of the corpus callosum is intimately adherent to the fornix. So close indeed is this adhesion that the separation is always attended with injury to the fornix. The deep-seated fibres of the corpus callosum seem to unite the lateral halves of the fornix.

The horizontal portion of the fornix, as exposed by this dissection, has the form of a triangle, the apex of which is directed forwards, and corresponds to the posterior angle of the septum lucidum. Its base is situate behind, and is enclosed by the posterior folded portion of the corpus callosum. The apex is prolonged into two rounded cords of fibrous matter, which pass downwards and outwards, in a somewhat curved course, with their convexity directed forwards. These are the *anterior pillars* of the fornix. As they descend, they diverge from each other. We can follow them down to the base of the brain, where they join two small tubercles, the *corpora mamillaria*, from which fibres are continued upwards and outwards into the substance of the optic thalamus.

The *posterior* pillars of the fornix are expansions of fibrous matter which are continuous with the angles of the base of its horizontal portion. These bands are continued into the lateral ventricle, and expand partly over the posterior horn, and partly over the hippocampus major in the inferior horn. The portion of the fornix which is thus continued into the inferior horn presents a fine concave edge directed inwards, which is the *corpus fimbriatum*.

It would thus appear that the fornix consists of a horizontal triangular portion *(corpus fornicis)* resting on four pillars, which take somewhat of a curved course, and form numerous connections with deep-seated and important portions of the brain. The anterior pillars are closely connected with the optic thalamus, with the tuber cinereum, and with the white matter which forms the floor of the ventricle. The posterior pillars are in intimate union with the posterior and middle lobes of the brain.

The fibres of the fornix are distinctly longitudinal. So that, supposing it to be commissural in its office, it may be stated to connect the anterior lobe of the brain and the optic thalamus with the posterior and middle lobes.

The fornix is divisible into two equal and symmetrical portions, one belonging to each cerebral hemisphere. These portions are united, as has been already stated, by the deep-seated transverse fibres of the corpus callosum, and by the terminal fibres of its posterior reflexion, which form, on the inferior surface of the fornix, a peculiar appearance called *the lyra*. The transverse white fibres stand out in relief, crossing at right angles the proper fibres of the fornix. In many subjects, however, this appearance is but faintly indicated.

The horizontal portion of the fornix rests upon a triangular process of pia mater, which is introduced into the interior of the brain, at the fissure beneath the posterior reflexion of the corpus callosum. This process is the *velum interpositum* already described at page 28.

The anterior pillars of the fornix bound in front a space in which the velum interpositum and choroid plexuses unite, and through which the lateral ventricles communicate with each other. This is the *foramen commune anterius*, described by the first Monro.* If a probe be laid transversely in this orifice, it will have above it the anterior extremity of the fornix, in front the anterior pillars, and behind it the point of junction of the three processes of pia mater.

Of the third ventricle.—If the fornix be divided transversely at about its middle, and the segments reflected, and if the velum interpositum be removed, a fissure, *the third ventricle*, is exposed, situate between the optic thalami. This fissure extends forwards between the anterior pillars of the fornix, where it is limited by a band of white matter visible without dissection in that interval. That band is *the anterior commissure*, which lies just in front of, and as a tangent to the convex border of the anterior pillars of the fornix.

At its posterior extremity the third ventricle becomes very much contracted in all its dimensions, and is continuous with a canal which leads to the fourth ventricle *(iter a tertio ad quartum ventriculum, Aqueductus Sylvii)*. The orifice of this canal is apparent at the posterior extremity of the third ventricle, and is bounded superiorly by a transverse cord of white matter, *the posterior commissure*, which extends for a short

---

* But previously recognised and described by Vieussens.

distance into the cerebral matter on either side. The base of the pineal gland rests upon this commissure.

In this stage of the dissection, a general view of the third ventricle is gained. This cavity evidently results from the apposition of the lateral halves of the brain proper, the parts which more immediately correspond being the inner surfaces of the optic thalami. The depth of the ventricle corresponds, in a great degree, to that of these bodies; but it manifestly increases towards the anterior extremity. Its floor is formed by a layer of grey matter continued from one side to the other, of the same nature as that which has been already described as covering the thalami. The deepest part of the ventricle is an infundibuliform depression, from which the tubular process, seen at the base of the brain (*fig.* 22, *l*), is continued down to the pituitary body. Just beyond this part is the anterior extremity of the ventricle, situate between the anterior pillars of the fornix and behind the anterior commissure; the depth of which is much less than that of the infundibulum.

The floor of the third ventricle corresponds to several parts of interest which have been enumerated along the middle of the base of the brain. Corresponding to the posterior extremity of the ventricle is the interval between the crura cerebri, the *pons Tarini,* or interpeduncular space. Next in order, in the direction from behind forwards, are the *corpora mamillaria,* which are succeeded by the *tuber cinereum* and *commissure of the optic nerves.* The anterior extremity of the ventricle corresponds to that portion of the tuber cinereum which extends between the optic commissure and the anterior reflection of the corpus callosum.

The roof of the third ventricle is formed by the velum interpositum, already described as giving support to the horizontal portion of the fornix.

The direction of the long axis of the third ventricle is obliquely downwards and backwards. Its anterior extremity being on a higher plane than its posterior, is therefore likewise superior.

Pineal gland.—We may here conveniently notice the position and connections of the *pineal gland*. This body, rendered famous by the vague theory of Des Cartes, which viewed it as the chief source of nervous power, is placed just behind the third ventricle, resting in a superficial groove which passes along the median line between the corpora quadrigemina. It is heart-shaped, and of a grey colour. Its apex is directed backwards and downwards, and its base forwards and upwards. A process from the deep layer of the velum interpositum envelopes it and serves to retain it in its place. From each angle of its base there passes off a band of white matter which adheres to the inner surface of each optic thalamus. These processes serve to connect the pineal body to the optic thalami. They are called *the peduncles of the pineal gland,* also *habenæ*. In general they are two in number, one for each optic thalamus. They may be traced forwards as far as the anterior pillars of the fornix. Posteriorly these processes are connected along the median line by some white fibres which adhere to the base of the pineal gland, as well as to the posterior commissure beneath, and which seem to form part of the system of fibres belonging to that commissure. A pair of small bands sometimes pass off from these fibres, along the optic thalami, parallel to the peduncles above described.

It appears, then, that the pineal gland has no other connexion with the brain than that which these *habenæ* or peduncles secure for it; otherwise this body might more appropriately be regarded as an appendage to the pia mater, in which it is involved in a kind of special capsule, from which it derives support and nutrition.

Grains of sand, similar in every respect to those previously described (p. 29) as connected with the internal processes of the pia mater, are found in the pineal body in a large proportion of instances in the adult. They seem to be accumulated in a cavity which is situate towards its base. Hence Soemmering gave to this collection of sabulous matter the name *acervulus*. When, however, the sand is abundant, it may be found upon the surface as well as in the centre.

The anterior commissure.—In examining the third ventricle, a rounded cord of very pure white matter is seen through the interval which is left by the divergence of the anterior pillars of the fornix in their descent to the base of the brain. This band is transverse, and appears to form a tangent to the convex border of those pillars. It may be traced outwards on either side through the anterior extremities of the corpora striata into the white substance of the middle lobes of the brain. A very little dissection is required to expose this cord in its entire extent. It seems placed in a canal hollowed in the cerebral matter. When exposed, its surface is perfectly smooth, indicating that fibres do not pass from it to the wall of the canal in which it lies. Examined in its whole extent, it presents the form of a curve with anterior convexity, and becomes gradually flattened and expanded towards each extremity, its component fibres becoming divergent and mingling with the white substance of that portion of the brain.

This system of fibres possesses the characters of a commissure or bond of connection between symmetrical portions of the brain on either side of the median plane, as distinctly as the corpus callosum itself.

The soft commissure.—The cavity of the third ventricle is partly occupied by a lamina of a light grey matter, which extends between the optic thalami of opposite sides. It forms a transverse horizontal plane dividing the ventricle into two portions, one above, the other below it. Sometimes it is divided and disposed as two planes. There is but little power of cohesion between its particles, so that in the recent state the separation of the thalami in the necessary manipulations will frequently cause its rupture. Hence the adjunct "soft" is appropriately applied to it, and by its connecting the thalami of opposite sides, this structure may be ranked with the other commissures. It does not extend throughout the entire length of the ventricle: both its anterior and posterior margins are concave and leave an open space between each extremity of the ventricle.

Thus far our examination includes the topographical anatomy of the cerebrum proper. The pineal body, indeed, scarcely lies within the confines of that segment of the encephalon, but from its internal relation to the third ventricle and the optic thalami, it must be included in the description of those parts. This body rests on the upper surface of that segment of the brain which lies intermediate to the cerebrum, cerebellum, and medulla oblongata, namely, the *mesocephale*. And we shall now proceed to a brief notice of this part and its connection to the other segments.

The mesocephale.—Four eminences are seen immediately behind the third ventricle. A transverse furrow separates them into an anterior and a posterior pair, and a longitudinal furrow along the median line divides the right and left pair from each other. The pineal body rests in the anterior extremity of the longitudinal depression. The anterior pair have been long named the *nates*, the posterior the *testes*. In the human subject the former are the larger. In the inferior mammalia these bodies are much more highly developed than in man, and exhibit a more marked difference of size.

The posterior of the corpora quadrigemina are apparently connected to the cerebellum by two columns of white matter, one of which passes into the central white substance of each cerebellar hemisphere. These are the *processus cerebelli ad testes*. They enter into the formation of the crura cerebelli. Each of them forms the superior layer of the crus cerebelli of its own side.

The interval between the *processus cerebelli ad testes* is occupied by a horizontal stratum of nervous matter composed of a thin layer of grey and of white matter. This is called the *valve of Vieussens*, although there is evidently nothing valvular in its nature or office. Its surface is marked by slight transverse depressions and eminences. The median lobe of the cerebellum overlaps and conceals it from view.

The valve of Vieussens* must be regarded as a portion of the median lobe of the cerebellum, which is extended forwards between the processus cerebelli ad testes. Its constitution is precisely the same as the laminæ of that body, and the transverse markings upon its superior surface are indications of imperfectly developed fissures between the rudimentary laminæ of which it is composed.

The corpora quadrigemina form the anterior superior part of the mesocephale. They lie above the crura cerebri, upon those columns of nervous matter by which the latter bodies are connected with the medulla oblongata. These columns are continuous above with the optic thalami, and below with the central portion of the medulla oblongata, *the olivary columns*, or

---

* *Valvula cerebri major* is the name which Vieussens applied to this process. He describes it as "membrana quam transversus medullaris tractus circa anteriora subit, processui vermiformi anteriori, processibus a cerebello ad testes et posticæ pontis Varolii parti adhæret et unitur." He further adds, " illam valvulæ vices gerere asserimus. Ex quo fit, ut habitâ officii et magnitudinis illius ratione, ipsam valvulam cerebri majorem nominemus, ut eam a membranaceis ligamentis distinguamus, quæ intra longitudinalis et lateralium sinuum cavitates valvularum minorum vices supplent et munia præstant."— *Neurographia Universalis*, p. 76. Ed. Lugd. 1716.

fasciculi innominati of Cruveilbier. They are distinguished by their reddish grey colour and their close resemblance in point of structure to the optic thalami. In transverse section they appear as two columns, circular in outline, quite distinct from the surrounding greyish matter in which they seem imbedded (*fig.* 28, *i*).

The lower half of the thickness of the mesocephale is formed by transverse curved fibres with anterior convexity, which extend between the lateral lobes of the cerebellum, and of longitudinal fibres which interlace with the superior layers of those transverse fibres and cross them at right angles. The former constitute the pons Varolii, a great commissure between the hemispheres of the cerebellum; the latter are, in greater part at least, the fibres of the anterior pyramids of the medulla oblongata, which ascend through the pons, and enter into the formation of the inferior layer of each crus cerebri.

In examining the inferior surface of the mesocephale, the pons Varolii, we observe that a longitudinal groove extends along its middle from above downwards. In this lies the basilar artery. Above the anterior edge of the pons, the crura cerebri are seen emerging, and diverging from each other as they pass, to enter, *stalk-like*, into the inferior surface of the cerebral hemispheres. Beneath its posterior edge, the medulla oblongata is seen, its anterior and middle columns passing through the mesocephale to the crura cerebri. On each side the fibres of the pons pass off into each hemisphere of the cerebellum and form the inferior lamina of each crus of that organ.

The cerebellum.—Some account of the general disposition of the cerebellum will serve to conclude this brief review of the topography of the brain. The superior surface of this organ is a little above the level of the quadrigeminal bodies. It is smooth and slightly convex. The lamellæ of the cerebellum are visible upon it, but cannot be separated without removing the arachnoid and pia mater. A notch is seen, dividing the posterior edge into two equal portions, and a larger notch exists in front, at which the cerebellum forms its connection with the mesocephale. These notches denote a subdivision of the organ into two *lateral portions*, or *hemispheres*, and *a median portion*. The superior surface of the median portion is called the *superior vermiform process*; its anterior terminal laminæ form the valve of Vieussens. On the inferior surface the hemispheres of the cerebellum are much more convex than on the superior. The median portion too is somewhat differently arranged on its inferior surface; it consists of a series of laminæ, following a transverse direction; those in its centre are of greater transverse extent than those at either extremity, whence the appearance of a crucial figure results. This is *the inferior vermiform process*.

The posterior margin of the cerebellum is convex, and corresponds to the concave surface of the occipital bone, the falx cerebelli occupying the notch in its middle. Along the line of this margin, the pia mater sinks into a deep fissure, which takes a horizontal direction from behind forwards, and divides the cerebellum into a superior and inferior portion.

As the brain, removed from the cranium, lies with its base upwards, the medulla oblongata is seen between the lateral hemispheres of the cerebellum occupying a portion of the depression between them, in which is the inferior vermiform process (*fig.* 22).

The *fourth ventricle* is a lozenge-shaped cavity situated in the upper and posterior part of the medulla oblongata, and formed by the separation of its postero-lateral columns (*corpora restiformia*). The cerebellum contributes to inclose it above by means of the anterior laminæ of the superior vermiform process and the valve of Vieussens, and below and behind by the inferior vermiform process (*fig.* 26).

CHAPTER V.

Of the Medulla Oblongata.

We now proceed to the examination of the various segments of the encephalon, with a more special reference to the structure and physiological bearing of each. It may be here remarked that, while all the segments are intimately connected with each other and are therefore mutually dependent, there is much in their structure to justify the assumption that each is capable of exercising an independent function, which is, however, liable to be modified by the influence which any one, or all of the other segments may have upon it.

We begin with the description of this segment because of its immediate connection with the spinal cord, for it is plain, since this is the connecting link between that centre and the intracranial mass, that whatever influence the latter may exercise upon the former, must be conveyed or propagated by the medulla oblongata.

It is proper to notice that the term *medulla oblongata* has not been employed in a uniform sense by all anatomists. Willis and Vieussens comprehended under this title all the parts from the corpora striata and optic thalami (both included) down to the commencement of the spinal cord.* The same signification was adopted by the writers who immediately followed these great anatomists. Winslow considers the medulla oblongata as " one middle medullary basis common to both

---

* See the quotation from the English edition of Willis, at p. 135.

cerebrum and cerebellum, by the reciprocal continuity of their medullary substances."* The *crura*, or *pedunculi cerebri*, constitute its anterior part: these seem to be lost in the corpora striata, as Winslow states, and therefore they are looked upon as the peduncles of the cerebrum. Its posterior portion is called the *extremity* or *cauda* of the medulla oblongata (*queue de la moelle allongée*). It is to this latter portion that Haller restricted the term medulla oblongata, and most modern anatomists follow his example. Rolando, however, still applies the term in its more extended sense.

In the present work, I adopt the phraseology of Haller as far as regards the term *medulla oblongata*. It seems to form an upper enlarged portion of the medulla spinalis, to which it stands in somewhat the same relation as the capital to the shaft of a column. Its superior limit is indicated by the posterior edge of the pons Varolii; its inferior is denoted by a horizontal plane extended between the occipital foramen and the first vertebra. A more natural line of demarcation, however, between this part and the medulla spinalis may be found in certain decussating fibres which are seen crossing the anterior median fissure of the former at its inferior extremity. No such limit as this, however, is found on the posterior surface (*fig.* 23).

The medulla oblongata has somewhat of a conical shape, its base being situate above at the posterior margin of the pons. It is slightly flattened on both anterior and posterior surfaces, more so on the latter than on the former.

The medulla oblongata admits of the same primary subdivision as the medulla spinalis, namely, into two equal and symmetrical portions separated from each other by an anterior and a posterior median fissure. The former is wide but not

* Winslow's Anatomy, translated by Douglas, vol. ii. p. 316. Edin. 1763.

THE MEDULLA OBLONGATA. 163

of great depth. It is occupied by a fold of pia mater. Its floor is formed by a layer of fibrous matter which has the same cribriform appearance as that of the anterior spinal fissure. These fibres are commissural, connecting the two portions of the medulla oblongata. The posterior fissure is very deep and

Fig. 23.

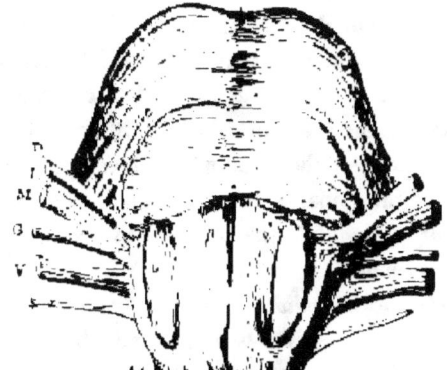

Anterior view of the medulla oblongata and pons Varolii.
(After Arnold.)

a, anterior extremity of the pons.
p, anterior pyramids.
d, decussating fibres of anterior pyramids.
o, olivary bodies.
A, arciform fibres.
D, portio dura.
I, portio intermedia of Wrisberg } Seventh pair
M, portio mollis } of nerves.
G, glosso-pharyngeal nerve
V, par vagum } Eighth pair
S, spinal accessory } of nerves.

narrow. It is not limited in front by a grey commissure as the posterior spinal fissure is, but by the posterior surface of the white commissure just described. A single layer of the pia mater passes into it. The continuity of the anterior fissure of the medulla oblongata and of that of the spinal cord is interrupted by the decussating fibres of the pyramids, (*fig.* 23, *d*,) but the posterior fissures are distinctly continuous with each other.

On either side of the median plane there are indications on the surface of the medulla oblongata, which suggest a subdivision of each half of the organ into four columns of nervous matter, through the medium of which it forms its connection with certain parts of the cerebrum and cerèbellum on the one hand, and of the spinal cord on the other. These columns are the *anterior pyramidal*, the *olivary*, the *restiform*, and the *posterior pyramidal*.

The anterior pyramidal columns, or *anterior pyramids*, (*figs.* 23, 24, 25, *p*,) are two prismatic bundles of fibrous matter which extend between the antero-lateral columns of the spinal cord and the lateral hemispheres of the brain. In the medulla oblongata each of these columns forms a compact body, which, when cut transversely, exhibits a triangular outline in its central portion, but that of a cylinder at either extremity. Each pyramid is limited on the outside by a superficial groove, which separates it from the olivary column, and on the inside by the anterior median fissure. Superiorly the pyramids pass into the mesocephale above the inferior fibres of the pons Varolii, and interlace with other fibres of the same system which occupy a more elevated plane. In its passage into the mesocephale, each pyramid experiences a marked constriction, which alters its form from a prism to a cylinder. The fibres, however, soon diverge and expand. As they ascend through the mesocephale they are crossed by the transverse fibres of the pons, and some grey matter occupies the interstices between them, with which it is probable that other fibres are connected, and are added to those of the pyramids, as they emerge from the mesocephale at its anterior extremity.

Fig. 24.

Anterior view of the medulla oblongata, shewing the decussation of the pyramids, and of the upper part of the spinal cord. (After Mayo.)
p, anterior pyramids. o, olivary bodies.
r, restiform bodies. d, decussating fibres.
al, antero-lateral column of the spinal cord.
c, anterior fissure of the cord, the floor of which forms the anterior commissure.

The pyramids gradually diminish in size towards the inferior extremity of the medulla oblongata. And here three sets of fibres may be distinctly noticed. The first, or *decussating fibres*, are the most numerous; they pass downwards and backwards into the antero-lateral column of the spinal cord on the opposite side, so that the right pyramid sends fibres into the left half of the cord, and the left pyramid into the right half of the cord. These decussating fibres consist of from three to five bundles from each pyramid, which in their descent cross and interlace with each other (*figs*. 24, 25, *d*). They differ in distinctness as well as in number in various subjects. The point at which the decussation takes place is about ten lines below the margin of the pons Varolii, and the interruption to the fissure, occasioned by the crossing of the fibres, occupies a space of from two to four lines. To expose these fibres clearly it is necessary to remove the pia mater carefully from the anterior surface of the medulla oblongata to some distance below the decussation, and it is, in general, of advantage to the preparation to place it in alcohol immediately after the removal of the pia mater.

A second set of fibres, very few in number, are continued from the pyramids directly down to the anterior surface of the cord on the same side, and appear to be continuous with some of the superficial fibres of the antero-lateral column. These fibres may be regarded as the direct channel of communication of each half of the medulla oblongata with the corresponding half of the spinal cord (*fig*. 25, *n*).

The third series of fibres vary considerably in point of developement in different individuals. They pass between the pyramids and the postero-lateral columns of the medulla oblongata, the restiform columns. They form a series of curves with their concavities directed upwards (*fig*. 23, A), crossing beneath the inferior extremity of the olivary body, and sometimes extending over a considerable portion of its surface. I have on several occasions seen these fibres so largely developed as to cover nearly the whole surface of each olivary body. These fibres are appropriately distinguished by the name *arciform* from their arched course (*processus arciformes*, Santorini).

When these fibres are so numerous as to cover the surface of the olivary body, we may observe that those which are nearest the margin of the pons Varolii are the least curved, and in some rare instances the uppermost ones exhibit no more curvature than the posterior fibres of the pons.

Fig. 25.

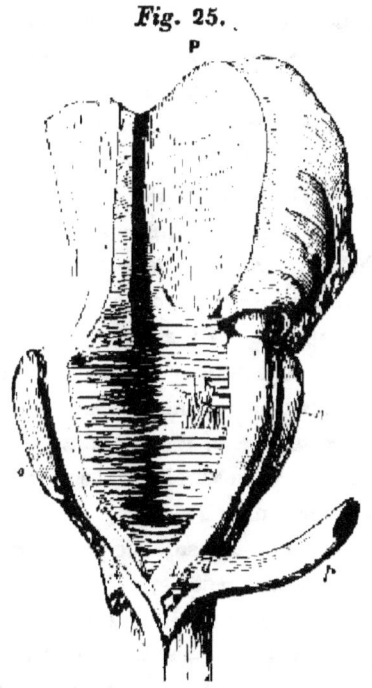

Anterior surface of the medulla oblongata, with a portion of the spinal cord and of the pons Varolii, as seen obliquely from the right side. (After Mayo.)

P, pons Varolii—its left half.
o, o, olivary bodies.
p, part of the right anterior pyramid, cut across near the inferior edge of the pons, and torn down, showing the passage of some of its fibres over to the left side and backwards.
d, decussating fasciculus of fibres of right pyramid.
d', decussating fasciculus of left pyramid.
n, non-decussating fibres of the right pyramid.

Both Santorini and Rolando have figured these fibres in their most highly developed state. The delineation given by the latter author, whilst it serves admirably as a diagram to show the general relation of the fibres, represents them as more numerous and distinct than I have ever had an opportunity of seeing them, and likewise exhibits them as passing upwards through the pons. This is certainly not the case. These fibres appear to incorporate themselves with the restiform bodies which connect the medulla oblongata with the cerebellum.

It seems to me that the arciform fibres may be properly regarded as a part of the same system as those which form the pons Varolii. They are largely developed in some quadrupeds, although they assume a different form. The fibres which constitute what Treviranus called *the trapezium*, appear to answer the same purpose as the arciform fibres; but, by reason of the non-developement of the olivary bodies on the exterior of the medulla oblongata, they do not take the curved course, which characterizes them in the human subject. These fibres cross the anterior surface of the medulla oblongata parallel to but distinct from the pons Varolii. They connect the pyramids and restiform bodies on each side.

By their continuation upwards the pyramidal bodies form a connection with the mesocephale, and also with the hemispheres of the brain through the medium principally of the corpora striata, and perhaps also of the optic thalami. Through the decussation of fibres which takes place just before the pyramids sink into the spinal cord, each cerebral hemisphere is connected with that half of the spinal cord which belongs to the opposite side of the body. By this arrangement is explained the influence which cerebral disease exercises upon the side of the body opposite to that on which it occurs. If the right hemisphere be irritated, convulsions are produced on the left side; if the right hemisphere be compressed, the left arm and leg and side of the face will be paralysed. So constant is this " crossed " influence of cerebral lesion that it can be attributed only to some uniform physical condition of the nervous centres. And that the anatomical disposition on which it depends is situate at the lower part of the medulla oblongata is proved, not only by the existence of these decussating fibres at this situation, but likewise by facts revealed by the phenomena of disease, and the results of experiment. Morbid lesions, for example, which have their seat *above* the decussation are, with rare exceptions, accompanied by affection of the opposite half of the body—those which involve the nervous centre *below* the decussation affect the body on the same side. Mechanical injury to the brain or spinal cord produces like effects. And so constantly is this the case that when we meet a case of paralysis or of convulsion affecting only one side, we confidently predict that the lesion on which it depends will be found on the opposite side of the brain.

This law of cerebral action has been known from the earliest periods of medical science, but the anatomical explanation of it, the suggestion of which dates as far back as the time of Aretæus,* has been generally admitted only within a comparatively recent period. This explanation was founded on the hypothesis of a decussation of fibres in the medulla oblongata to a greater or less extent. Santorini, indeed, laid it down that decussation took place not only in the lower part of the medulla oblongata, but likewise at the anterior and posterior margins of the pons Varolii.† But it is quite impossible, by

* Περι αιτιων και σημειων χρονιων παθων, βιβλ. A, κεφ. ζ, p. 87, Ed. Kuhn.

† Santorini must have been well acquainted with the decussating fibres of the pyramids, which he clearly describes. The whole passage is worth being quoted here. " Id autem triplici potissimum in loco animadvertere potuimus; in utraque scilicet priore, posterioreque annularis protuberantiæ crepidine atque maxime in imo medullaris caudicis quâ in spinalem abit. In priore itaque annularis

our ordinary means of observation, to detect any such connection between the anterior pyramids elsewhere than at their inferior extremity. In many instances I have thought that the fibres of the commissure which forms the floor of the anterior fissure presented an appearance as if decussation took place along the entire length of the pyramids. But the numerous foramina by which that commissure is penetrated to give passage to vessels for the central substance of the medulla, are very apt to give rise to a fallacious appearance of this kind.

It has been stated that there are exceptions to this law of cerebral action. Such certainly must be extremely rare, for in the course of a considerable experience for many years I have not met with an unequivocal instance in which paralysis occurred on the same side with cerebral lesion. The analysis which Burdach has given of 268 cases of paralysis in which there was lesion of a single hemisphere, shows very strikingly how rare must such an exception be. Of these cases he states that 10 were accompanied with paralysis of both sides, and

protuberantiæ parte, quâ superius reflexa pro comprehendendis oblongatæ medullæ cruribus in anguli formam interiùs producta tenuatur, sic ex concurrentibus fibris, strictioriqùe agmine coeuntibus altera alteram scandit ut præter mirum implexum decussatio luculentissimé appareat. Idipsum fermé in postica ipsius crepidine occurrit. Eo iterum in loco, qui quarto ventriculo subjicitur, præter varios fibrarum ordines et colores, in adversum latus productas et decussatas fibras commodé spectavimus. Si ea tamen evidenter uspiam conspicitur, profectò quam evidentissimè duas vix lineas infra pyramidalia atque adeo olivaria corpora conspici potest. Quà enim in longitudinem producta linea seu rimula pyramidalia corpora discernuntur, si leniter deducantur, probè prius eo potissimum loco arctissimè hærente tenui meninge nudata, non tenues decussari fibrillas, sed validos earundem fasciculos in adversa contendere, quam apertissimè demonstrabunt." Observ. Anat. cap. iii. § xii. p. 61. Ed. Lugd. Bat. 1739.

that 258 had hemiplegia. And of the hemiplegic cases, the paralysis occurred on the same side as the cerebral lesion in only 15.

The full explanation of these exceptions has yet to be discovered. The anatomical connection of each anterior pyramid with the spinal cord, however, affords some clue to it. This takes place, it will be remembered, not by the decussating fibres only, but by straight and perpendicular ones also; so that each pyramid is connected with both halves of the spinal cord, first and principally with the opposite half; and, secondly, and by much fewer fibres, with that of the same side. When those parts of the brain are affected, with which the decussating fibres are connected, the paralysis will be crossed; when, on the other hand, the direct fibres are engaged, the paralytic affection will occur on the same side of the body as that on which the lesion has occurred. But even on this explanation it is difficult to understand how these latter cases should be of such rare occurrence, and still more, how hemiplegia is so frequently accompanied with a perfect state of sensation and motion on the other side. In the present state of our knowledge, however, this is the only contribution which anatomy can offer towards the determination of this difficult question.

Of the restiform columns.—The lateral, and, in great part, the posterior portion of the medulla oblongata is formed on each side by a thick and rounded *rope-like* column, called the *corpus restiforme*. It is composed chiefly of fibrous matter, and its constituent fibres take a longitudinal direction. There is no line of demarcation between them and the fibres of the spinal cord, with the antero-lateral and posterior columns of which they seem to be continuous. Traced upwards, the restiform columns pass a little outwards, and by their divergence contribute to the increased size of the medulla oblongata at its base.

To see the connections of these columns completely, the posterior surface of the medulla oblongata should be examined. The restiform columns form the greater part of this surface. They increase in thickness as they ascend. Their outer margin forms a gentle curve, which is concave. Their inner border is connected in its lower portion to two small bands of fibrous matter, between which the posterior median fissure is situate; these are the *posterior pyramids* (Y, *fig.* 26). In its upper portion, the inner border of each restiform column is free, and forms the outer boundary of a lozenge-shaped depression, *the fourth ventricle.* Whilst the connection of the cerebellum with the posterior surface of the medulla oblongata is undisturbed, the exact relation of these bodies to the ventricle cannot be seen. It is necessary to raise up the inferior portion of the median lobe of the cerebellum, to expose the cavity of the ventricle; or this may be effected by dividing the median lobe along the middle line.

Each restiform column ascends to the hemisphere of the cerebellum of the corresponding side. The whole of its fibres appear to penetrate that organ, and contribute to the formation of its crus, the middle layer or peduncle of which it forms. This is very well shown in the analytical figure at p. 128, (*fig.* 20,) where *r* is the restiform column passing upwards and outwards into the hemisphere of the cerebellum.

The distinction between the restiform and olivary bodies on the surface is indicated by the line of origin of the eighth pair of nerves, which may be said to emerge along the anterior margin of the former. A narrow band of fibres, very distinct in some brains, occupies the depression between the posterior edge of the prominent olivary body and the line of emergence of these nerves. This band has been well delineated by Rolando, Reid, and others; it probably forms a part of the

Fig. 26.

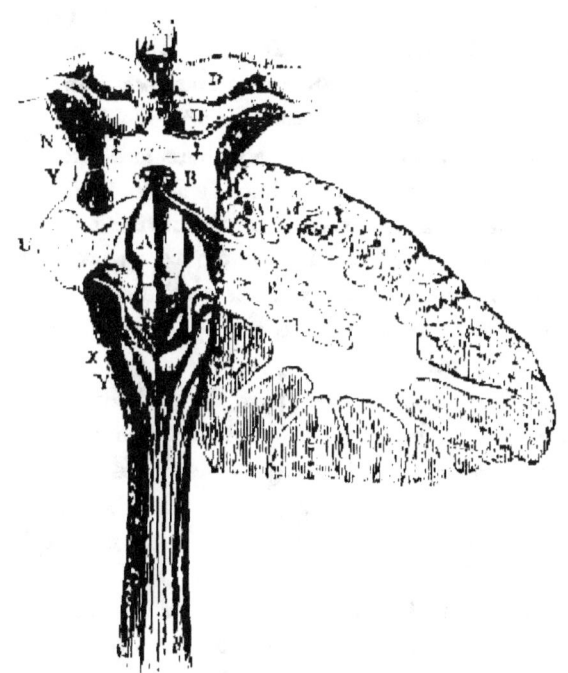

Posterior view of the medulla oblongata, with mesocephale and part of cerebellum of an infant. (*After Foville.*)

S, pineal gland.
D, nates.
D', testes.
+ +, points of emergence of fourth pair of nerves.
Y, posterior pyramids.
X, restiform columns.
A, F, floor of the fourth ventricle, formed by the olivary columns, the fissure between which is the calamus scriptorius.
Y', posterior surface of mesocephale.
B, valve of Vieussens.
N, anterior surface of crus cerebri.
R, corpus dentatum or rhomboideum.

cerebral fibres of the medulla oblongata, and ascends through the pons.

The direction of the fibres of the restiform columns is longitudinal. Those which are situate most posteriorly pass directly downwards, and are distinctly continuous with the posterior columns of the spinal cord. The fibres which form the lateral and anterior part of the restiform columns pass downwards and forwards to the antero-lateral columns of the cord. A superficial groove, varying very much in distinctness in different subjects, which passes upwards from the line of emergence of the posterior roots of the spinal nerves, indicates the distinction of these two sets of fibres. If the posterior column be separated from the antero-lateral in the spinal cord, the separation may be easily carried upwards along this line, in a specimen which has been sufficiently hardened.

From the description now given, the restiform columns may be regarded as the connecting fibres between the cerebellum and the spinal cord. They may be designated the *cerebellar fibres* of the medulla oblongata in contradistinction to the others, which are entirely connected with the mesocephale and with the cerebrum.

Rolando describes the restiform body as containing grey matter—the *grey tubercle* of Rolando. This grey matter, however, may be more correctly regarded as a portion of the central nucleus of the medulla, from which very probably some fibres of the restiform body emerge.

The posterior pyramidal columns.—On each side of the posterior fissure we find a narrow column, sufficiently distinct from the restiform columns. These may be traced downwards through the cervical region of the cord, and even into the dorsal or lumbar, according to Foville. They taper gradually to a fine point, the situation of which varies in different subjects. Superiorly they form the inferior and part of the lateral boundary of the fourth ventricle. Their innermost fibres end abruptly in a blunt extremity, whilst the external ones are continued upwards on each side of the ventricle (*fig.* 26, Y).

Olivary bodies.—The oval bodies, which form a relief upon the surface of the medulla oblongata, have been long known by the names *corpora olivaria, olivæ*. They occupy the interval between the anterior pyramids and the restiform bodies, separated, however, from the latter by the narrow band of fibrous matter above described.

The surface of each olivary body is crossed to a greater or less extent by the arciform fibres, as already described. Sometimes it is necessary to remove these fibres, in order to expose the proper texture of the olives.

The superficial layer of each olivary body is evidently fibrous, and the constituent fibres seem to take a longitudinal course. If a section be made so as to remove the prominent convexity of this body, it will be seen that the white matter of which it principally consists encloses a layer of vesicular or grey matter disposed in a peculiar manner. This grey layer presents the appearance of a waving line enclosing white matter. If the section of the olivary body be made transversely, the grey waving line is still present, but it presents a convex border outwards, and is open within, being evidently continuous with the central and less definitely disposed grey matter of the medulla. And when the section is vertical, and so as to divide the olivary body in its entire length, the convex border of the grey line is still external, but it is open towards the interior of the medulla.

This grey layer, contained within the olivary body, is called the *corpus dentatum* (*corps festonné*, Fr.) It is evidently a capsule of vesicular matter continuous below with that of the cord, internally with that of the central substance of the medulla oblongata, and superiorly with that of the mesocephale (*o, fig.* 27). Its disposition, in a convoluted form, has doubtless reference to the packing of a certain quantity of this matter into a given space, and to the important object of bringing the vesicular and fibrous matter into connection as extensively as possible.

Fig. 27.

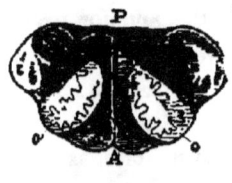

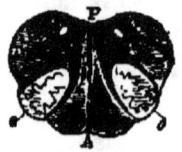

Transverse sections of the medulla oblongata.
A, anterior.   P, posterior.
o, olivary bodies, in which is seen the undulating line of grey matter which forms the corpus dentatum.

It has been very commonly supposed that the olivary bodies are mere gangliform masses laid upon certain ascending fibres of the medulla, and that they may be readily removed without injury to the deeper-seated parts. Either of the two following modes of dissection will, however, serve to point out the erroneousness of this view. If the anterior pyramids be removed, a concave surface is left between the two olivary bodies, in which their continuity with the central substance of the medulla is distinctly seen. This central substance, which forms a substratum on which the anterior pyramids rest, and from which it is not improbable that some of the fibres of the pyramids emerge, is of considerable density. Each olivary body appears gradually to merge into it; or, adopting another mode of description, it seems to protrude, forming a relief on the exterior, in the interval between the pyramidal and restiform bodies on each side. Or, a transverse section, as in *fig.* 27, will exhibit a similar continuity between the olivary bodies and the central substance of the medulla.

According to this view, then, the existence of the olivary bodies in the human brain and that of the Quadrumana indicates a high developement of the central substance of the medulla oblongata as compared with its other nervous columns. In all the vertebrate animals below man, the medulla oblongata increases with the bulk of the body, and like the spinal cord evidently bears a direct relation to it. This high developement appears, however, to affect more especially the restiform and pyramidal bodies, and their connecting fibres, *the trapezium*. The former do not leave any space between them, and the central columns do not extend to the surface; and from the absence of any excessive developement of grey matter, we find no such arrangement as that which gives rise to the *corpus dentatum* in man.

Olivary columns.—The union of the olivary bodies with the central part of the medulla oblongata constitutes what may be termed the *olivary columns*. These columns pass into the mesocephale, occupying a plane superior to that of the pyramidal fibres and of the transverse fibres. They may be traced upwards to the crus cerebri, where they seem to merge into the optic thalami, and to form a connection with the corpora quadrigemina posteriorly.

The olivary columns are seen distinctly in their ascent to the brain in the fourth ventricle, as two cylinders, (A, F, *fig.* 26,) which form the floor of that cavity. They are separated from each other by the longitudinal fissure which is continued upwards from the posterior fissure of the medulla oblongata.

In the fourth ventricle the olivary columns are crossed by the fibres of origin of the portio mollis of the seventh pair of nerves, the white colour of which in the recent specimen contrasts strikingly with the greyish hue of the columns themselves. We here see distinctly that these columns are the source of origin of these nerves, and no doubt they are equally so of all the nerves which are connected with the medulla oblongata, namely, the fifth pair, the eighth, the ninth, and probably also of the sixth.

The relation of the olivary columns in their upward course, to the other constituents of the mesocephale and crura cerebri, may be conveniently demonstrated in examining transverse sections of those parts. We shall, therefore, return to this subject in describing the anatomy of those portions of the brain.

The following interpretation of the various columns of the medulla oblongata, referred to in the preceding description, has much foundation in their anatomical relations.

The olivary or central columns constitute the fundamental part of the medulla oblongata; that, on which its action as a distinct and independent centre depends, and in which the proper nerves of this segment of the encephalon are implanted. The continuity of those columns with the optic thalami and corpora quadrigemina materially enhances their physiological influence, and denotes their intimate association with some of the most important functions of the brain. And it may be added, that this connection of the medulla oblongata with parts which are ordinarily described as pertaining to the brain itself, shews that the original application of the term by Willis and Vieussens to a much greater extent of the encephalon is certainly more consistent with the physiological anatomy than that which is now employed for the convenience of description. There can be no doubt that the extent of this central and fundamental portion of the nervous system is limited above by the optic thalami and below by the spinal cord.

The anterior pyramids connect the cerebral hemispheres with the spinal cord, the principal bundles of fibres decussating each other on the middle line, so that the right pyramid is the medium of connection by the greater number of its fibres between the right hemisphere of the brain and the left half of the cord, but by a much smaller number between that same hemisphere and the right half of the cord. And so also of the left, *mutatis mutandis*. It is highly probable too that the anterior pyramids derive fibres from the locus niger of the crus cerebri and the vesicular matter of the mesocephale. These fibres, therefore, connect those segments with the spinal cord, but whether they contribute to the formation of the decussating or non-decussating bundles, or to that of both, it is impossible to determine.

The restiform columns are evidently the connecting fibres between the hemispheres of the cerebellum and the posterior and antero-lateral columns of the spinal cord. And the posterior pyramids connect the posterior part of the medulla oblongata with the cervical and dorsal regions of the cord.

Nerves.—Numerous nerves are connected with the medulla oblongata—a fact which serves greatly to enhance its importance as a centre of nervous action. These nerves are the sixth pair, which are connected with the anterior pyramids just behind the posterior border of the pons; the ninth pair, or hypoglossal nerves, which emerge along the anterior border of the olivary body; the seventh pair (*portio mollis* and *portio dura*), which emerge just behind the upper extremity of the olivary body; and the eighth pair, which arise along the posterior margin of the olivary body.

CHAPTER VI.

Of the Mesocephale.

The pyramidal and olivary columns may be readily traced, as already explained, from the medulla oblongata up to the cerebral hemispheres; the former becoming united chiefly with the corpora striata, the latter with the optic thalami.

In that part of their course which is intermediate to the medulla oblongata these columns become mingled with certain transverse fibres, and with more or less of vesicular matter, and with them contribute to form a mass which is the connecting link between all the segments of the cerebellum, and may be compared to a railroad station, at which several lines meet and cross each other. This is the *mesocephale* or *mesencephale*. The name was suggested by Chaussier, inasmuch as it forms " to a certain extent the middle and central part of the encephalic organ, the bond which unites the several bundles of fibres which contribute to its formation."

The mesocephale may be isolated from the other segments by dividing the crura cerebri just beyond the anterior margin of the pons, and the crura cerebelli as they penetrate the hemispheres, and the medulla oblongata on a level with the posterior edge of the pons. The crura cerebri emerge from it in front: the medulla oblongata is connected with its posterior surface: on either side it is prolonged into a crus cerebelli. Its inferior surface, which is very convex and looks forwards, is composed of the thick layer of arched fibres which form the *pons Varolii*;

and on its superior surface, which looks backwards, are the corpora quadrigemina, the processus cerebelli ad testes, and part of the floor of the fourth ventricle (*fig.* 26).

According to Chaussier, its weight is equal to about the sixtieth or sixty-fifth part of the entire brain.

We shall describe separately the inferior and the superior surfaces of this segment of the encephalon, and its intimate structure as unfolded by sections.

The inferior surface, (*pons Varolii, annular protuberance,*) convex from side to side, is interrupted along the median plane from behind forwards by a shallow groove in which the basilar artery usually lies, giving off in its course numerous minute capillaries to the nervous structure of the mesocephale.

When the pia mater has been stripped off this surface, it is seen to be very evidently composed of a series of transverse fibres which take an arched course. The fibres are collected into large fascicles separated from each other by very distinct intervals, so that there is no part where the fibrous structure is more apparent than here. They form arcs of circles, not concentric, lying one behind the other in a series nearly parallel. Owing to this want of complete parallelism the width of this surface measured from before backwards is much less at each extremity than in the centre. The anterior margin is convex, and forms a thick edge crossing the crura cerebri like a bridge; hence the term *pons* was applied by Varolius to the whole series of fibres. The posterior border is concave, less curved than the anterior, and crosses the anterior pyramids and olivary columns, as the latter does the crura cerebri. The intervening fascicles of fibres become gradually less curved as they approach the posterior margin.

These transverse fibres form a stratum of considerable thickness at the inferior surface of the mesocephale. Some grey matter is deposited between the less superficial layers which constitute it. The most deeply-seated layers are penetrated and crossed at right angles by the ascending fibres of the anterior pyramids. A remarkable interlacement takes place at this situation between the vertical and transverse fibres—the latter passing alternately in front of and behind adjacent bundles of the former. Some of the vertical fibres seem to sink into and connect themselves with the grey matter.

A transverse vertical section of the mesocephale gives a more complete view of the exact extent of the transverse fibres. They are found to occupy rather more than one-third of the depth of the exposed surface. Their disposition in laminæ is very apparent. Those which are nearest the centre of the mesocephale have between them considerable intervals, which are filled up by vesicular matter, through which pass vertically the fibres of the pyramids. The intervals between the laminæ gradually diminish towards the inferior surface of the pons, and the quantity of intervening vesicular matter becomes proportionally less, and disappears altogether from between those laminæ the intervals of which are not traversed by the fibres of the pyramids.

The transverse fibres pass on either side into each hemisphere of the cerebellum, contributing with the *processus cerebelli ad testes* and the restiform bodies to form the *crura cerebelli*. They are the *inferior peduncles* of these crura.

The anatomy of these transverse fibres evidently denotes that they serve to connect the right and left cerebellar hemispheres, as *commissures*, and in a manner strikingly analogous to that in which the fibres of the corpus callosum connect the cerebral hemispheres. This view of the office of these fibres is strongly confirmed by the fact that their number is always in the direct ratio of the size of the lateral hemispheres, and that when the hemispheres are absent, these fibres no longer exist. When, therefore, the cerebellum consists only of a median lobe, there is no pons Varolii.

Some of the transverse fibres nearer the inferior surface appear to dip in along the median line, and to pass upwards and backwards, forming a vertical plane of fibres which divides the mesocephale into two symmetrical portions, and Chaussier imagined that a decussation took place at this situation. The groove in which the basilar artery lies is formed partly by the greater condensation which is produced along the median plane by this arrangement, and partly by the slight bulging on either side of it, caused by the ascent of the anterior pyramids. These fibres are continuous with a series of similar ones in the medulla oblongata (*antero-posterior fibres* of Cruveilhier).*

The extent of the superior surface of the mesocephale may be limited in front by a line which passes from side to side just before the anterior of the corpora quadrigemina, and posteriorly by the base of the valve of Vieussens. This occupies a much greater space than the inferior surface. It is an inclined plane, and passes downwards and backwards, being concealed by the anterior laminæ of the superior vermiform process of the cerebellum and the posterior border of the corpus callosum.

The *corpora* or *tubercula quadrigemina* are four rounded eminences—gangliform bodies—disposed in pairs (*fig.* 26, D, D'). The anterior pair are larger than the posterior. The former have been distinguished as the *nates*, the latter the *testes*.† These bodies are situate further forwards than the pons,

---

* Sir C. Bell has given a good representation of these septa in his latest papers in the Philosophical Transactions.

† In reference to these absurd appellations Willis has the following remark : " Prominentia orbicularis—quarum usus longè nobilior videtur, quam ut viliora ista natium et testium nomina mereantur."

and are chiefly connected with the superior surface of each crus cerebri.

The nates are of a deeper grey colour than the testes. Both pairs are similar in structure to the optic thalami. When cut into, they appear to consist of fibrous matter intermingled with vesicular. Thin sections examined with the microscope exhibit intricate interlacements of tubular fibres with vesicular matter interposed—a true ganglionic structure.

An important fact deserves special notice as indicating that vesicular matter is found in these bodies in considerable quantity. The pia mater which adheres to their surface abounds in minute bloodvessels, and in separating it these are seen to penetrate the tubercles in vast numbers. This layer of pia mater contributes to form the velum interpositum.

The quadrigeminal bodies are the analogues of the optic lobes in birds, reptiles, and fishes. In these classes there is only a single pair of tubercles. They are of considerable size in birds, and form a conspicuous portion of their encephalon. The division into four takes place only in Mammalia. The anterior are the larger in herbivorous animals, the posterior in the Carnivora. In most quadrupeds these bodies are concealed from view by the posterior lobes of the brain; but in Rodentia they are exposed in consequence of the imperfect developement of the brain in the backward direction.

The quadrigeminal bodies rest upon two processes of fibrous matter, which extend backwards to the median lobe of the cerebellum, and forwards to the optic thalami. These processes form a connection between the thalami and the quadrigeminal bodies and the cerebellum. They have been variously designated *processus cerebelli ad testes, processus cerebelli ad corpora quadrigemina, processus cerebelli ad cerebrum.*

The *valve of Vieussens* intervenes between these processes, and closes the fourth ventricle at its upper part.

A longitudinal groove separates the right and left pair of quadrigeminal bodies. The anterior extremity of this groove forms an expanded and somewhat flattened surface on which rests the pineal gland (*fig.* 26, S). From the posterior extremity a small band extends to the valve of Vieussens, called *frænum*. An incision made vertically downwards along the course of this groove exposes the canal through which the fourth ventricle and the third communicate (*iter a tertio ad quartum ventriculum*). This canal communicates with the posterior part of the third ventricle by an opening which is situate beneath the posterior commissure, and with the superior extremity of the fourth ventricle beneath the valve of Vieussens.

The fourth pair of nerves are seen upon this surface, attaching themselves to the processus cerebelli ad testes, or to the Vieussenian valve, or to the posterior pair of quadrigeminal bodies.

Besides the anterior pyramids, the olivary columns are continued through the mesocephale to form with the former the crura cerebri. These columns are exposed along the floor of the fourth ventricle; higher up, however, they are surrounded by a lightish grey matter, form the superior stratum of each crus cerebri, separated from the quadrigeminal tubercles by the processus cerebelli, and finally merge into the optic thalami. Their course is well displayed in *fig.* 20, where *f* represents the olivary columns, *t* the processus cerebelli ad testes, and *v* the pons penetrated by *p*, the pyramids.

The olivary columns retain their greyish hue in their upward course. Their cylindrical form is very apparent on the floor of the fourth ventricle; but it is still more obvious on viewing a transverse section, when each olivary column appears as a cylinder, to be distinguished from the rest by its roundness and its peculiar colour.

No other mode of dissection conveys so much knowledge

of the anatomy of this part as a transverse section, carried from above downwards through either pair of quadrigeminal bodies, and inclined a little backwards so as to pass through the pons, or forwards to pass through the crura cerebri. The parts which may be observed on such a section, enumerated from above downwards,—are, 1, either pair of quadrigeminal tubercles; 2, between and beneath them, the iter cut across; 3, on either side of this, fibrous matter; 4, below this on each side, the section of each olivary column; 5, planes of transverse fibres interlacing with longitudinal ones, and vesicular matter between the planes; 6, transverse fibres forming the pons Varolii.

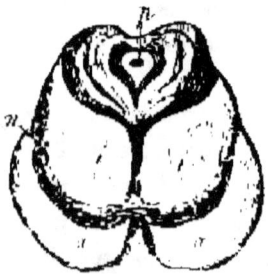

Fig. 28.

Plan of a transverse vertical section of the mesocephale anterior to the pons, passing through the crura cerebri.

p, iter a tertio ad quartum ventriculum. This is surmounted by a pair of the quadrigeminal tubercles.
i i, olivary columns.
n, locus niger.
a, the inferior plane of fibres diverging upwards, continuous with the anterior pyramids.

From the preceding description of the mesocephale it may be concluded that two classes of elements enter into its formation. These are *intrinsic* and *extrinsic*. The former consists of the masses of vesicular matter, with which the fibrous matter, whatever be its course, is intimately connected. Such are the grey matter of the quadrigeminal bodies; that light grey matter which surrounds the olivary columns in their upward course; the darker matter which intervenes between the transverse fibrous lamellæ; and more in front, that which forms the locus niger of the crus cerebri.

The *extrinsic* elements are those which pass through this segment, being continuous with some portion of a neighbouring segment, or serving to connect the grey matter of the mesocephale with the hemispheres of the cerebrum or cerebellum, or with the medulla oblongata. The fibres which form the inferior layer of the pons are perhaps the only element that does not connect itself in some way with the grey matter of the mesocephale, since they seem simply to pass across from one crus cerebelli to the other. The deeper transverse fibres, the pyramids, the olivary columns, the processus cerebelli ad testes, all connect neighbouring parts with the intrinsic matter of the mesocephale.

It is plain, then, that anatomy affords abundant grounds for the conclusion, that the mesocephale must be regarded as a distinct centre, connected by numerous bonds of union with the other segments of the brain.

If further proof of this were wanting, it would be found in the connection of two important nerves with this segment. These are the fifth and the fourth pairs. The former penetrate between the superficial fibres of the pons which spread out upon the crus cerebelli; the latter are connected with the superior surface of the mesocephale.

CHAPTER VII.

Of the Cerebellum.—The fourth ventricle.

This remarkable portion of the encephalon, so called from its general resemblance to the cerebrum, of which it is, as it were, the diminutive, is situate behind the mesocephale and medulla oblongata. It is lodged in a compartment of the cranium, the floor of which is formed by the fossæ of the occipital bone, and which is separated from the cavity occupied by the cerebrum, by the horizontal process of the dura mater, previously described as the *tentorium cerebelli.* This process forms a partition between the inferior surface of the posterior lobes and the superior surface of the cerebellum.

The cerebellum, like the cerebrum, is at its highest point of developement in the human subject. It exists as a very distinct portion of the encephalon in all the classes of vertebrate animals, and exhibits a marked gradation of increase from Fishes, through Reptiles and Birds, up to Mammals.

In Fishes and Reptiles it consists of a single lobe, overhanging the posterior surface of the medulla oblongata, and closing the fourth ventricle partially like a valve. It is, in general in these classes, smooth on its surface, and exhibits no complication of structure, no subdivision into laminæ. But in the sharks a manifest increase in size and an incipient lamellar arrangement are distinctly observable, which shew that in them this organ is more highly developed than in any other fishes.

In birds a similar complication of structure takes place to

a much greater extent, and a lateral lobe or appendage is added on each side to the single central organ which constitutes the cerebellum of fishes and reptiles. And in the mammiferous series, the lateral lobes along with the central portion experience a progressive augmentation of size (proportionally to the body), and a corresponding complexity of structure up to the quadrumana and man.

The best and most obvious subdivision of the human cerebellum is into the *median lobe* and the *lateral lobes* or *hemispheres*. The former is the fundamental and primitive portion of the organ; the latter, although each exceeds the median lobe in size, and therefore they conjointly form far the largest portion of the cerebellum, are appendages, which in man assume great physiological importance. The median lobe has likewise been called *vermiform process*, the upper and lower laminæ being distinguished as the *superior* and *inferior vermiform processes*.

From the tables already given it would appear that the cerebrum is to the cerebellum in the proportion of 8 or 9 to 1 in the adult, and in the infant, according to Chaussier, as 16 or 18 to 1. The average weight of the cerebellum is, according to Professor Reid's researches, 5 oz. 4 dr. in the male, and 4 oz. 12 dr. in the female.

The cerebellum seems to keep pace, in its developement, with the cerebrum. It attains its greatest size, both in male and female, at the same age as the cerebrum. At the most advanced ages, however, it seems to diminish with greater rapidity than that organ.

Some variety appears to occur as regards the relative developement of cerebellum to cerebrum in the adult. Chaussier remarks that he had in some instances found the cerebellum equal to a seventh or a sixth part of the weight of the cerebrum, but rarely the eleventh or twelfth.

There does not appear to be any good grounds for the assertion that the cerebellum is more developed in proportion to the brain in the female than in the male. Professor Reid's extensive series of researches show, beyond all question, that it maintains the same proportionate bulk in both sexes.

It has also been asserted that castration, or disease of the genital organs, such as would destroy the generative instinct, causes wasting of the cerebellum. If both testicles be removed, the whole cerebellum, it is said, degenerates; if only one, the hemisphere of the opposite side is affected.

The most complete refutation of this assertion is afforded by M. Leuret's series of observations of the brains of geldings and entire horses. These researches, indeed, shew that in stallions the cerebellum is proportionally smaller than in mares or geldings, and that in geldings it is larger than in mares. It is very evident from them that mutilation of the sexual organs does not cause degeneration of the cerebellum.

The shape of the cerebellum is that of " an ellipsoid flattened from above downwards."* Its principal diameter, which is transverse, is from three-and-a-half to four inches in length; the antero-posterior diameter is from two inches to two inches and a half; the anterior part is about two inches in thickness; whilst near its posterior edge it does not measure above half an inch.

At its anterior edge the cerebellum is notched, and receives fibres by which it is connected to the cerebrum and mesocephale. This notch is of considerable transverse extent, and is semilunar in shape. The greater portion of the posterior part of the mesocephale corresponds to it. By Reil this is called the *semilunar fissure*. In it we find several parts which the anatomist should study; namely, on the highest plane, the

* Cruveilhier.

Fig. 29.

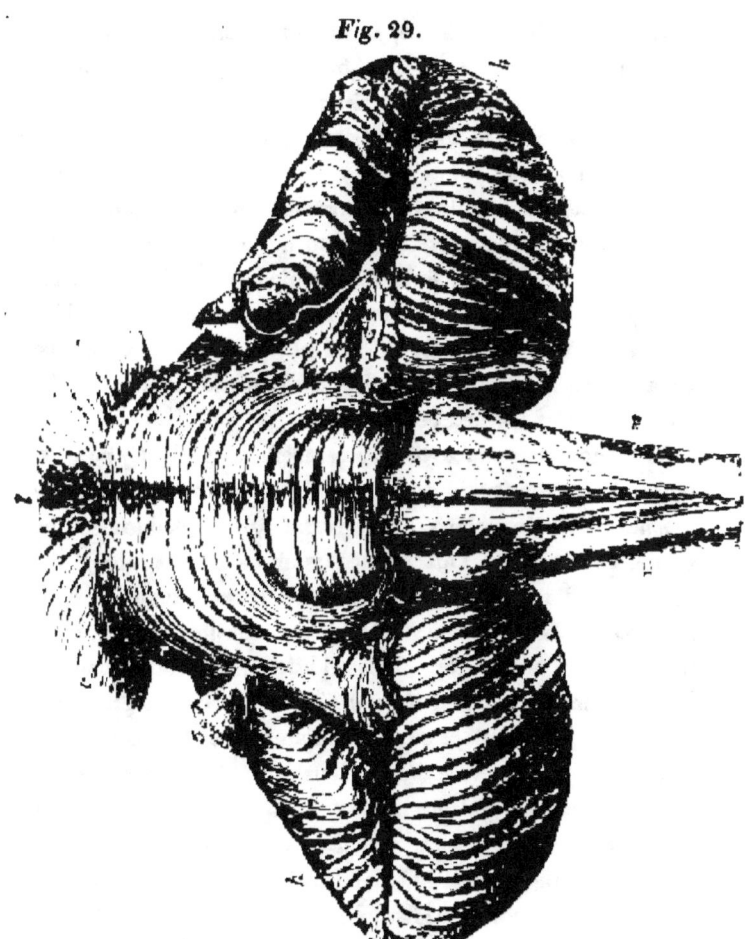

Front view of the cerebellum, with medulla oblongata and mesocephale.
(After Foville.)

c, c, crura cerebri.
l, locus perforatus, or pons Tarini.
m, m, medulla oblongata.
h, h, semilunar fissure.
n, the flock on the right side, or lobule of the vagus.
5, fifth nerve. On the left side a layer appears to be extended from this nerve which contributes to form the crus cerebelli.

processus cerebelli ad testes, separated by the valve of Vieussens, and, beneath these, the fibres of the restiform columns, and the right and left extremities of the pons Varolii, all of which combine to form the crus cerebelli or central stem of each lateral lobe.

The posterior margin is interrupted in its middle by a vertical notch, which divides it into a right and left portion. This notch is wider in front than behind, whence Reil called it the *purse-like fissure ;* the term *posterior notch* is preferable. It receives the falx cerebelli, and at its bottom we observe a continuity between the superior and inferior laminæ of the median lobe of the cerebellum.

The superior surface of the cerebellum is slightly convex, inclined backwards and downwards. It terminates in front by a concave margin, which overlaps the parts contained in the semilunar fissure. This surface is more convex along the middle than on either side. In the latter situations it is inclined and nearly plane; but in the former it resembles more the surface of a cylinder. This middle portion corresponds to what is commonly called *the superior vermiform process*: it is in fact the upper surface of the median lobe of the cerebellum.

On its inferior surface the subdivision of the cerebellum into two symmetrical portions is very apparent, by reason of the existence of a deep fissure which proceeds from before backwards along the median line, and is continuous behind with the posterior notch. This fissure is called the *valley (vallecula,* Haller; *grande scissure mediane du cervelet)*. It separates the hemispheres of the cerebellum, each of which presents a very convex surface, corresponding to each occipital fossa. The arachnoid membrane is extended from one to the other, towards the posterior part of the fissure, leaving a considerable space between it and the pia mater, which is traversed by some fine bundles of fibrous tissue and occupied by subarachnoid fluid.

This space has already been referred to as the *posterior conflux* of Majendie.

The anterior part of this fissure receives the upper and posterior portion of the medulla oblongata. The remainder of it is occupied by the inferior surface of the median lobe of the cerebellum, presenting a remarkable cruciform arrangement, which will be presently described.

Another very remarkable fissure requires a special notice. It is horizontal, and passes into the substance of the cerebellum, dividing it into an upper and an inferior portion. By inserting the handle of a knife along the posterior margin of the cerebellum, this fissure may be shewn to pass forwards to a considerable depth, and to communicate on each side with the semilunar fissure, whilst it is interrupted in the middle posteriorly, by the notch. Its inner surface is lined by a process of pia mater, which sinks into it.

The right and left cerebellar hemispheres exhibit a general symmetry, which is, however, not always perfect, as a manifest difference is sometimes observable in their sizes. And a corresponding want of symmetry may be frequently seen in the right and left fossæ of the occipital bone.

Both the hemispheres and the median lobe are composed of an assemblage of laminæ closely applied to each other. Each lamina consists of a thin layer of white or fibrous matter, between two of grey or vesicular substance, which are continuous along the outer margin of the former. Thus the exterior of the cerebellum consists of a stratum of vesicular matter, which forms a cortex to the enclosed white or fibrous substance. The laminæ are separated from each other by fissures, and are covered by pia mater, which adheres closely to them, and penetrates to the floors of the fissures.

The laminæ are collected into sets on the superior as well as on the inferior surface. Each set forms a lobe. Each lobe is surrounded by a *deep* fissure, which separates it from the next adjacent lobes.

It is necessary to distinguish the fissures which separate the laminæ from those by which the lobes are bounded. The former are very shallow: the latter are deep, and penetrate quite to the central stem of the hemisphere.

By removing the pia mater carefully from the surface of the hemispheres, and from the deep fissures, the shape and boundaries of the lobes may be clearly demonstrated. Or if a vertical section of a hemisphere be made, the deep fissures may be readily distinguished from the superficial ones which separate the laminæ; and in this way also the lobes may be demonstrated.

The floor of each deep fissure is formed by white matter. And as the deep fissures intervene between the lobes, laminæ of the lobes constitute their walls, and the superficial fissures which separate these laminæ open into them.

On the superior surface of the cerebellum two principal lobes may be distinguished. These are *the square lobe* and the *posterior superior* lobe, according to the nomenclature of Reil, whose descriptions cannot be surpassed in minuteness or accuracy. (*Fig.* 30, A, P.)

The anterior margin of the square lobe overhangs the semilunar fissure; its posterior margin is a little behind the level of the floor of the posterior notch. By careful separation of its laminæ or by a vertical section, it may be shewn to consist of eight lobules, each having a stem of fibrous matter derived from the central one of the hemisphere.

The posterior superior lobe (P, *fig.* 30,) forms the posterior part of the superior surface of the cerebellum; its posterior margin is that of the hemisphere; the horizontal fissure separates it from the posterior inferior lobe. It is separated from its fellow of the opposite side by the posterior notch.

On the inferior surface of each hemisphere the following lobes are readily distinguishable. (*Fig.* 31.) We enumerate them, passing from before backwards.

1. *The amygdala,* so called from its resemblance to an enlarged tonsil (*a, fig.* 31). This and its fellow of the opposite side form the lateral boundaries of the anterior extremity of the valley, and are in great part covered by the medulla oblongata.

2. Behind the amygdala is the *biventral lobe,* wedge-shaped, narrow towards the valley, wide towards the semilunar fissure (*b, fig.* 31). Its laminæ are curved with their concavity forwards and inwards, and it is united with its fellow of the opposite side by laminæ which cross the valley forming part of the inferior vermiform process.

3. The *slender* lobe, which consists of a few laminæ curved parallel to the posterior ones of the biventral lobe (*c, fig.* 31).

4. The *inferior and posterior* lobe, which extends to the posterior edge of the hemisphere. The inner margin of each of these lobes constitutes the lateral boundaries of the posterior notch.

Such is the constant disposition of the superior and inferior surfaces of the cerebellum. A defect of symmetry is sometimes produced by the inequality of corresponding lobes; but those above enumerated are always present. So definite an arrangement must obviously have some physiological import. What that may be it is impossible even to conjecture, and we must be, for the present, content with a concise statement of the facts of the anatomy. Much analogy exists between this arrangement and that of the convolutions on the surface of the brain, many of which exhibit a constancy of position and form quite as remarkable.

The median portion of the cerebellum is also composed of laminæ, which are continuous with those of the hemispheres, but their arrangement on the superior and inferior surfaces is so different as to demand a separate description. On the superior surface the laminæ are separated from each other by fissures, in the same way as those which constitute the hemi-

Fig. 30.

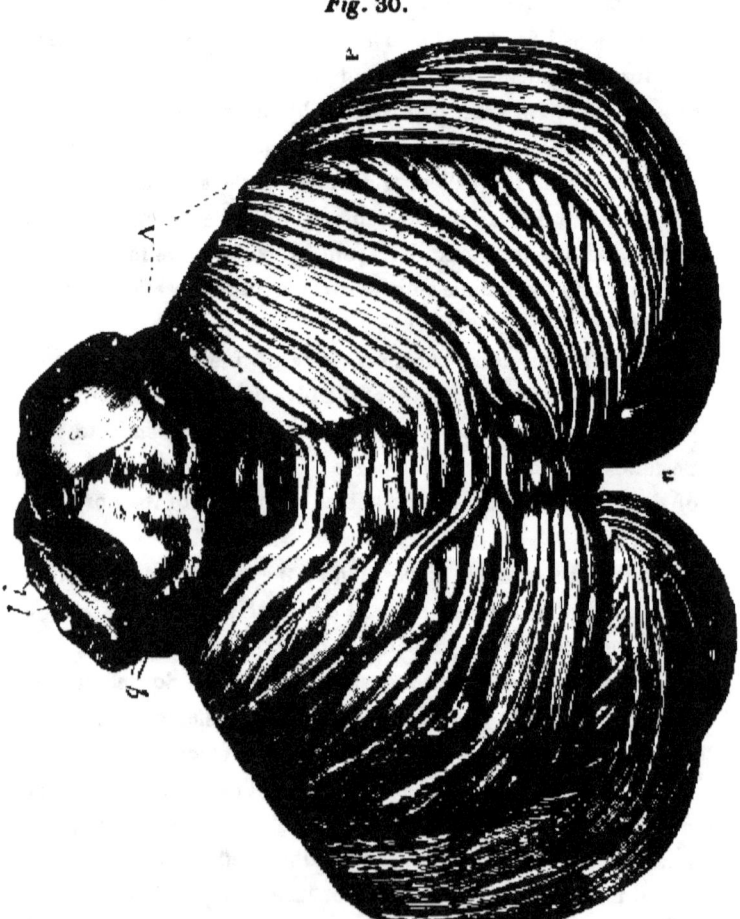

Superior surface of the cerebellum.
A, the square lobe ; P, the posterior superior lobe ; S, superior layer of the crus cerebri ; V, superior vermiform process ; q, tubercula quadrigemina ; locus niger ; i, inferior layer of the crus cerebri.

spheres, and they are collected into sets forming lobes which correspond to and connect those of the lateral hemispheres. These laminæ are curved, their anterior margin being very slightly convex (*fig.* 30). The edges of these laminæ, as they lie in close apposition, resemble the segments or rings of a worm; whence the term *vermiform* has been applied to this as well as the inferior surface of the median lobe. The laminæ take for the most part a vertical direction, with the exception of the anterior and posterior ones, which pass gradually to the horizontal, the free margins of the former being directed forwards and those of the latter backwards. The posterior laminæ form the floor of the posterior notch: the anterior form, by their adhesion to each other, the layer known by the name of *valve of Vieussens*, which fills up the interval between the *processus cerebelli ad testes*.

The laminæ which form the superior surface of the median lobe, (or the superior vermiform process,) are considerably fewer than those of the hemispheres. This explains the less depth of the median lobe, when measured from before backwards, than of the hemispheres. Two or more of the laminæ of the latter are united to a single lamina of the former, and thus the superior vermiform process serves as a transverse commissure to the superior laminæ of the hemispheres.

The inferior surface of the median lobe, or *inferior vermiform process*, is likewise composed of laminæ, which take a transverse direction and present a free convex border, with some resemblance to the rings of a worm in action. (*Fig.* 31.) These laminæ are not all of equal transverse extent. The middle and posterior are the broadest; the anterior gradually diminish in size. Hence the body which results from the conjunction of all the laminæ has a triangular form, its apex being anterior and its base posterior, corresponding to the notch between the hemispheres. The laminæ which occupy its middle have a greater depth than the rest, and give to the body a greater prominence at this situation.

Certain deep fissures divide the inferior vermiform process into segments which evidently correspond with and connect the lobes into which the hemispheres are subdivided on their inferior surface.

These segments may be very readily distinguished from each other, and the names which the accurate Reil has given them are sufficiently appropriate. By separating each segment from the adjacent ones and tracing its lateral relations, the anatomist may form a better idea than by any other means of the way in which this portion of the cerebellum is connected with the hemispheres.

The anterior extremity of the inferior vermiform process projects into the cavity of the fourth ventricle, and serves to close it at its inferior extremity. It is a pointed process, furrowed transversely, continuous by its base with the rest of the vermiform process. Reil has named it *the Nodule* (*n*, *fig*. 31). From either side of it a valve-like membrane (*V, V, fig*. 31), of exquisite delicacy, extends forwards and outwards towards a lobule which is attached to each crus cerebelli near to the origin of the auditory nerve. These membranes resemble very much in shape the semilunar valves of the aorta. By their attached margin they adhere to the crus cerebelli, and their free margin projects into the cavity of the fourth ventricle. Their inner extremities adhere to the nodule, and are connected to each other by a thin membrane of precisely similar texture, which is a commissure to them. Reil gives to the two membranes and their intermediate connecting one the name of *posterior medullary velum*.* The lateral membranes were first described by Tarin and Malacarne. When the fourth ventricle

---

* The valve of Vieussens is the anterior medullary velum.

has been carefully opened in a recent cerebellum, it is very easy to demonstrate them by passing the handle of a knife under them.

The structure of these lateral wings of the inferior medullary velum is readily ascertained. Their delicacy is such that they admit of being examined by the microscope without pressure or other manipulation. They consist of tubular fibres of various sizes, taking a transverse direction, that, namely, of the long diameter of each wing, covered by a layer of nucleus-like particles as an epithelium. They seem to connect the nodule to the small lobules of the pneumo-gastric nerve above mentioned (the *flocks* of Reil), or to connect those lobules themselves as a commissure.*

The nodule pushes before it, into the fourth ventricle, a fold of the pia mater, connected with which on either side are several small granulations, or *Pacchionian bodies*. It is called *the choroid plexus* of the fourth ventricle. We can easily trace it to be continuous with the pia mater which covers the lobules of the seventh pair of nerves.

Next to the nodule, below and behind it, is a small lobe, called by Reil the *spigot (Zapfen)*, (*s, fig.* 31,) with a pointed extremity directed downwards and forwards. It consists of several small laminæ separated by their fissures. Behind it is a larger lobule, which forms the most prominent portion of the inferior vermiform process, called by Reil, from its form, the *pyramid* (P, *fig.* 31). Its apex is directed downwards and backwards, and it likewise consists of numerous small laminæ.

These lobules of the inferior portion of the median lobe serve to connect others of the lateral hemispheres. The *spigot* connects the almond-like lobes; the *pyramid* the biventral and the slender lobes.

* Although it does not appear that Reil used the microscope, his statement respecting the structure of these wings is perfectly correct.

Posterior to the pyramid are a series of laminæ which extend to the posterior notch and form its floor. These pass directly from one side to the other, their free margin being convex and directed backwards. They connect the posterior inferior lobes. And some of the most anterior of them, which do not project to the surface, connect the slender lobes as well as some of the anterior laminæ of the posterior inferior lobes. These latter laminæ of the inferior vermiform process, Reïl distinguishes by the name of *long* and *hidden commissure (langen verdeckten Commissur)*, and the former constitute his *short and exposed commissure (Kurzen und sichtbaren Commissur)*.

Above the last-named commissure is a single lamina which forms a line of demarcation between the inferior and the superior vermiform processes, serving to connect the upper and posterior lobes of the hemispheres. This is *the single commissure (einfache quer Commissur)*.

It will serve to elucidate the foregoing necessarily intricate description, if I sum up with the following enumeration of the lobes of the hemispheres, specifying at the same time the commissures by which they are connected, i. e. the lobes of the superior and inferior vermiform processes which serve that purpose.

1. On the superior surface of the hemispheres.

a. The square lobes, consisting of eight lobules, which are connected by as many, or nearly so, of the superior vermiform process.

b. The upper and posterior lobes, connected by the single commissure, to be sought for on the floor of the posterior notch.

2. On the inferior surface of the hemispheres.

a. The amygdalæ, united by the spigot.
b. The biventral lobes.
c. The slender lobes.

Fig. 31.

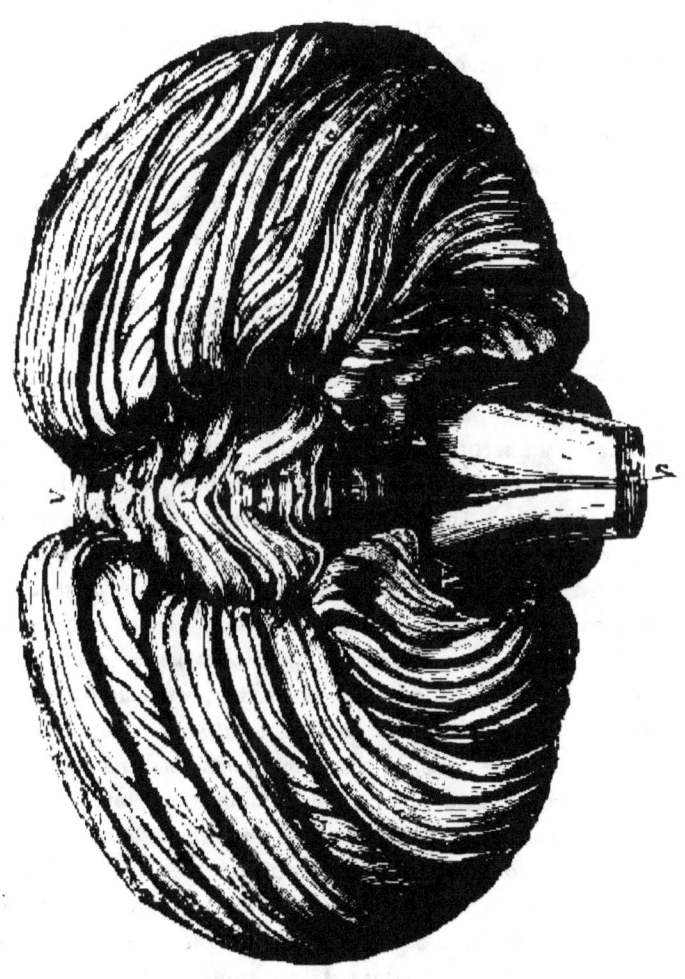

Inferior surface of the cerebellum.

V, inferior vermiform process ; *p*, posterior pyramids ; *r*, restiform bodies ; *a*, amygdalæ ; *b*, biventral lobe ; *c*, slender lobes ; P, pyramid ; *n*, nodule ; *V, V*, inferior medullary velum.

The biventral lobes and the anterior laminæ of the slender lobes are united by the pyramid.

d. *The posterior inferior lobes,* connected by the short and exposed and the long and hidden commissures.

The flocks or lobules of the pneumogastric nerve, (*lobule of the auditory nerve,* Foville,) which are situate altogether anterior to the hemispheres and attached to each crus, are united by the posterior medullary velum, and through it appear to have some connection with the most anterior portion of the inferior vermiform process.

A vertical section of either hemisphere of the cerebellum or of its median lobe displays its structure, and serves further to demonstrate the subdivision into lobes above described. When either hemisphere is cut in the vertical direction, the surface of the section displays a beautiful ramification of fibrous matter, the smaller branches of which are enveloped by laminæ of grey matter. This appearance has such a resemblance to the trunk of a tree with its boughs and branches, that it early received and has continued to retain the name of *arbor vitæ.* The trunk of the tree is represented by a central nucleus of white matter, from the upper and lower surfaces of which branch off, some at a right, others at an acute angle, several laminæ, each of which forms the parent stem of a number of other branches. Each of the primary branches is the foundation or central stem of a lobule. Laminæ of fibrous matter are seen branching from both sides of it immediately after its separation from the nucleus. Sometimes the primary branch bifurcates, and each division of it forms the stem of what may be called a sub-lobule. The ultimate branchings are covered by a layer of grey matter. If we suppose that one of the primary branches is composed of a certain number of laminæ of fibrous matter, the secondary ramifications from it will equal them in number. In most instances these secondary branches subdivide into two or more tertiary ones, which, as well as the branch from which they spring, are enclosed in grey matter. (*Figs.* 20, 26.)

A vertical section of the median lobe gives quite a similar appearance to that of the hemispheres. The central nucleus breaks up into primary branches, which become the centre of the lobules of which it consists. (*Figs.* 26, 33.)

The ramifications of the central nucleus, whether of the median lobe or of the hemispheres, separate from it only in the vertical plane or from before backwards; in the latter direction, however, to a very slight extent. Hence these branches are directed only upwards, or downwards, or backwards. The fibrous matter of the median lobe is continuous, without any line of demarcation, with that of the hemispheric lobules. By reason of this disposition of the fibrous matter, the surface which is exposed by a horizontal section through the entire cerebellum, presents a very different appearance from that which results from a vertical section. It consists of a plane of fibrous matter bounded on the sides and behind by a narrow cortex of grey matter.

The white matter consists exclusively of fibres, chiefly of the tubular kind and of all degrees of size. These, in the more distant ramifications, penetrate the vesicular matter of their grey cortex, and form some unknown connection with its elements. The grey matter consists of three layers, readily distinguishable by the naked eye from their difference of colour. The external layer is the darkest, and consists chiefly of granular and vesicular matter. The next or intermediate layer is of a light colour, and is composed of a stratum of fine nucleus-like particles. The third layer has the greatest thickness, and is immediately in contact with the fibrous matter; it is intermediate in point of colour to the other two, and consists of numerous vesicles of the caudate kind, especially with branching processes and nerve-tubes of all sizes. The dark colour of the external layer is doubtless owing in a great measure to the great numbers of capillary vessels which enter it; the greater paleness of the inner stratum is to be attributed to the intermixture of the white fibres, whilst the light colour of the middle stratum is intrinsic. From the usual dependent position of the cerebellum in the dead body, it always appears to contain more blood than the cerebrum.

Corpus dentatum.—If, in making a vertical section of either hemisphere of the cerebellum, the incision be made so as to leave two-thirds of the hemisphere on its outside, a peculiarity will be observed on the surface of the section which deserves a separate consideration. The central white nucleus is interrupted by a very remarkable undulating line of vesicular matter, which is convex towards the posterior margin of the hemisphere, but open in front towards the crus cerebelli.

This constitutes the *corpus dentatum* or *rhomboideum* of the cerebellum. It presents a remarkable resemblance to the structure of the same name which is met with in the olivary body of the medulla oblongata. It is evidently a capsule of vesicular matter which is enclosed in the inner third of the central white nucleus of the cerebellar hemisphere, being nearer its superior than its inferior surface. The peculiar undulating arrangement of it doubtless has reference to the accommodation of a certain extent of surface in a limited space. The fibrous matter enclosed by it seems derived from the processus cerebelli and from the restiform body.

The central stem of fibrous matter to which the several lobules, both of the hemispheres and the median lobe of the cerebellum, adhere, (*crus cerebelli,*) is formed by three bundles of fibres, each situate on a different plane. These are the *peduncles* of the *crus cerebelli*. Through them the cerebellum forms a connection with other parts of the encephalon.

The superior layer or peduncle is a bundle of fibres which extends to the corpora quadrigemina, and may be traced beneath them to the optic thalami. These are the *processus cerebelli ad testes*, but from their being obviously a medium of connection between the cerebellum and the cerebrum, they may be better named *cerebro-cerebellar commissures*. It is worthy of remark, that these are the only fibres which appear to connect these two segments of the brain. The middle layer is continuous with the restiform bodies, *processus cerebelli ad medullam oblongatam*. And the inferior layer is evidently derived from the transverse fibres of the pons Varolii, which thus pass from one hemisphere to the other, and constitute a great commissure to the cerebellar hemispheres. These fibres, moreover, connect each hemisphere to the mesocephale (*fig.* 20, *t, r, v*).

From this triple constitution of the crus cerebelli, it is plain that the cerebellum may exert an influence upon, or be affected by the optic thalami or quadrigeminal bodies, the restiform columns, or the mesocephale.

Of the fourth ventricle.—This is a rhomboidal cavity, situated at the upper and posterior part of the medulla oblongata, and extending over part of the superior surface of the mesocephale. It is limited superiorly by the posterior margin of the testes, and inferiorly by the superior blunt extremity of the posterior pyramids. Its two lateral angles correspond to the entrance of the restiform bodies into the crura cerebelli. In fact, it is formed by the divergence of the restiform columns in their ascent to the hemispheres of the cerebellum. The median lobe of the cerebellum lies over the fourth ventricle, and conceals it from view. The anterior lobule of the inferior vermiform process, *the nodule*, projects into it, and closes it below. On either side of this lobule a process of pia mater, with small granulations upon it, is found. These processes are the *choroid plexuses* of the fourth ventricle. Around these and thence on to the nodule, the proper membrane of the ventricle is reflected, and thus its cavity is shut out from any communication with the subarachnoid cavity. A vertical section in the median plane, or a little to one side of it, displays this arrangement well. (*Fig.* 32.)

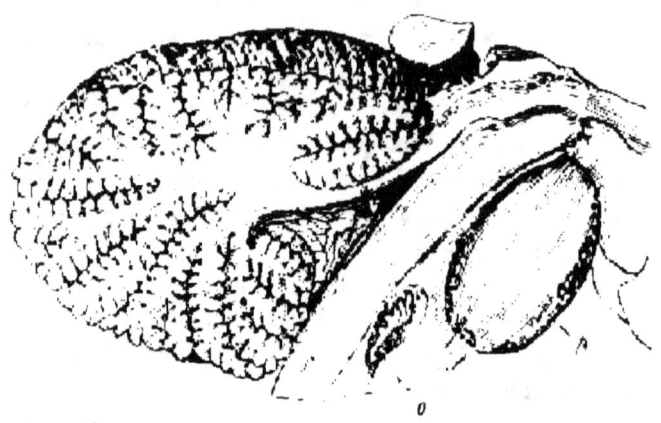

Vertical section of the median lobe of cerebellum, mesocephale, and medulla oblongata, to shew the fourth ventricle.

o, corpus dentatum; f, posterior surface of medulla oblongata; p, pons Varolii; a, processus cerebelli ad testes; v, cavity of the fourth ventricle.

Along the floor of the fourth ventricle we find the central or olivary columns of the medulla oblongata extending upwards to the optic thalami. A fissure, continuous with the posterior median fissure, separates these columns, and terminates above in a canal which penetrates the mesocephale, to reach the third ventricle: *iter a tertio ad quartum ventriculum* or *aqueduct of Sylvius*. On either side of the fissure certain bundles of white fibres, continuous with the auditory nerves, join it at right angles, crossing over the olivary columns. This fissure, with its white fibres on each side, has been compared to a pen with its barbs, and hence called *calamus scriptorius*.

The fourth ventricle, although sometimes called the ventricle of the cerebellum, properly belongs to the medulla oblongata. It is present in all the vertebrate classes, and in size bears a direct proportion to that of the medulla itself.

CHAPTER VIII.

Of the hemispheres of the brain—The convolutions—Corpora striata—Thalami optici—Corpora mamillaria—The commissures—Tuber cinereum—Pituitary body—The ventricles.

A MASS of fibrous matter, covered on its exterior by a convoluted layer of vesicular matter, inflected towards the mesial plane above and below a pair of gangliform bodies, *(optic thalamus* and *corpus striatum,)* which it thus encloses in a cavity or ventricle—this, with certain fibres connecting its anterior to its posterior parts, forms a cerebral hemisphere. The hemispheres of opposite sides are applied to one another along the mesial plane, leaving the fissure-like interval called the third ventricle; and they are united by a plane of transverse fibres, the greater part of which is placed above that ventricle, but which bends down anteriorly as well as posteriorly, closing the fissure at those situations.

Of the convolutions.—That which first attracts attention in connection with the cerebral hemispheres, as affording the highest physiological as well as anatomical interest, is their convoluted surface. This can only be well displayed by stripping off the pia mater. The appearance which is then presented has been variously described by different writers. It has always seemed to me to resemble the folded surface formed by the mucous membrane of the stomach when the muscular coat is very much contracted. The rugæ of that membrane become enormously developed by the excessive contraction of the muscular coat: the mucous membrane not possessing any contractile power is thrown into thin folds to adapt it to the diminished capacity of the stomach. Its folded state indicates a great disproportion between the extent of the mucous surface and that of the muscular tunic. If both surfaces were equal, neither of them would be thrown into folds. In examining the surface called *centrum ovale,* which is exposed by a horizontal section through the hemisphere above the level of the corpus callosum, we obtain an explanation of the formation of the convoluted surface of the brain. That plane of fibrous matter is surrounded by an undulating margin of vesicular matter, the foldings of which give rise to the convoluted appearance of the cerebral surface. The fibrous matter is adapted to this irregular surface, not by any similar folding, but by the prolongation of its fibres into the concavities of the folds. It is only by means of these prolongations that an equality obtains between the surface of grey matter and that of fibrous matter which it covers. In brains devoid of convolutions, the vesicular and fibrous surfaces are applied to each other as two layers disposed in concentric circles. There are no irregularities in either one or the other. But any increase in the extent of the grey surface involves a corresponding complication in that of the fibrous matter, which is effected by the prolongation of the fibres at certain situations. Were we to suppose two brains in which the quantity of fibrous matter in the hemispheres was equal, the quantity of grey matter in one might be increased considerably, and therefore become convoluted without involving any other alteration in the fibrous matter than the elongation of certain bundles of fibres at particular situations.

The existence of convolutions on the surface of the hemispheres, as contrasted with the absence of them, indicates an increase in the developement of the dynamic matter. A convoluted brain, even although actually smaller than one with

Fig. 33.

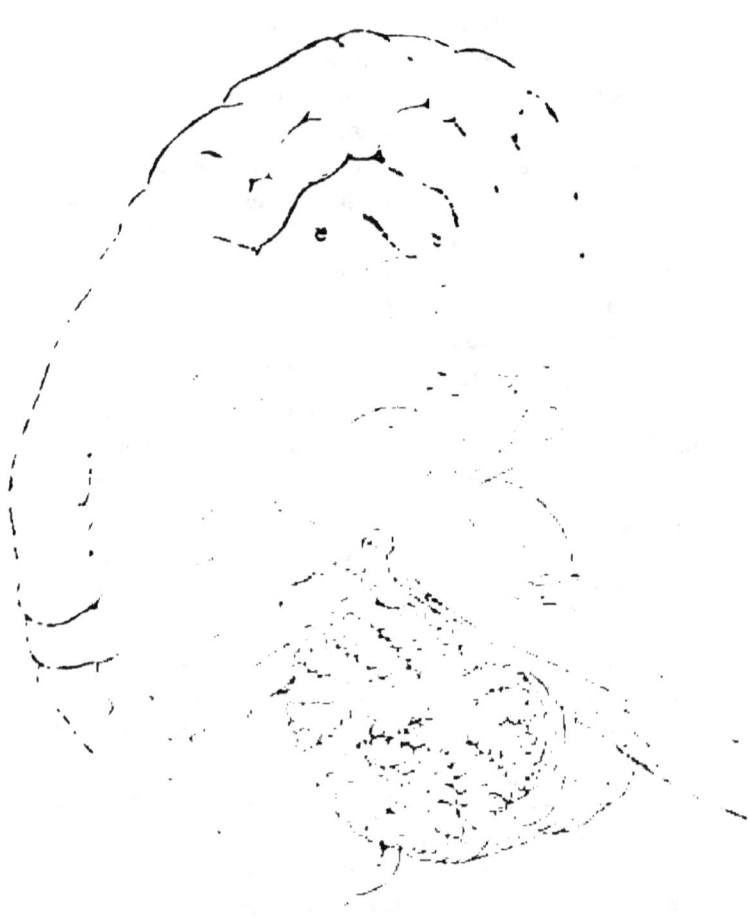

Vertical section of the adult human brain. (After Arnold.)

The position of the internal convolution with reference to the corpus callosum is well displayed. The median lobe of the cerebellum has been cut through, and the fourth ventricle exposed. *a, a, a,* internal convolution, (*d'ourlet,* Foville); *c,* corpus callosum; *o,* fornix; *n,* septum lucidum; *f,* pineal body; *i,* anterior commissure; *h,* hypophysis, or pituitary body; *t,* pons Varolii; II, second pair, or optic nerves; IV, fourth ventricle.

Fig. 34.

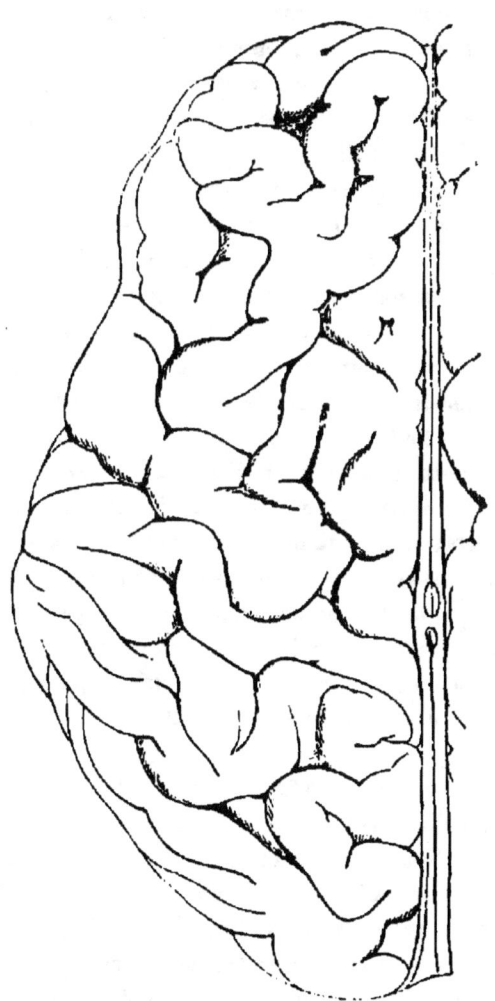

Superior surface of the right hemisphere of the adult human brain.

The undulating form of many of the convolutions is very well seen, and the general characters of the convoluted surface are displayed.

a smooth surface, would yet indicate a higher degree of mental power, inasmuch as it possesses a larger quantity of the vesicular matter relatively to its fibrous matter.

Cerebral convolutions are wanting in all the classes below Mammalia. They are likewise absent from the brains of many animals of the families Rodentia, Cheiroptera, Insectivora, some of the Marsupialia, and Monotremata. The brains of these Mammalia resemble very closely, as regards the characters of the cerebral hemispheres, the brain of Birds. There is not a trace of a convolution upon them, and the only fissure is the imperfectly developed one of Sylvius. The squirrel, the bat, the mole afford examples of brains deficient in convolutions. In some genera of the families Insectivora and Marsupialia, however, we find an approach to the convoluted cerebral surface in certain depressions marked on the exterior of each hemisphere. The fissure of Sylvius is more developed, and certain depressions, taking for the most part a longitudinal course, are seen on the surface of each hemisphere. The brains of the rabbit, the beaver, the guinea-pig, the agouti shew these fissures. They are generally regular in different individuals of the same genus, and they are symmetrical, i. e., of the same length and direction, and occupy the same place on each hemisphere.

Leuret remarks, in reference to the dogma of Gall and Spurzheim that the presence and number of the convolutions are in direct relation to the volume of the brain, that such is far from being universally the case; and I am glad to refer to so excellent an authority in confirmation of the view which I have advocated respecting the true signification of the cerebral convolutions. According to this anatomist, the ferret, which has several well-marked convolutions on each hemisphere, has a brain no larger than that of the squirrel, which is entirely devoid of them, and which has not even the few fissures which faintly indicate their first developement in the brains of the rabbit, the beaver, the agouti, &c. And the animals last named have the brain actually *larger* than that of the cat, the pole-cat, the roussette, *(Pteropus vulgaris,)* the unau, *(Bradypus didactylus,)* the sloth, *(Bradypus tridactylus,)* and the pangolin, all of which possess convolutions.

All mammiferous animals, excepting those mentioned in the preceding paragraphs, have convolutions which exhibit more or less of complication. This complication has evidently no connection with the general organization of the animal, inasmuch as we find animals, in the same family with those which possess numerous convolutions, exhibiting a very slight developement of them. The monkeys, the dolphin, the elephant, exhibit the most numerous convolutions of any of the Mammalia inferior to man, in whose brain the convoluted surface reaches its highest point.

Each fold on the surface of the brain is ordinarily called *a convolution*, whatever be its position, size, or direction. It consists of a fold of grey matter, enclosing a process of white or fibrous matter. On each side of it is a sulcus or groove, in which we find the same elements, a fold of grey or vesicular matter—concave externally, convex internally—the fibrous matter adhering to its convex surface. As the convolution exhibits no essential difference of structure from the sulcus, it is plain that the former portion of the brain's surface cannot differ in physiological office from the latter. We describe particular convolutions, not because they are to be regarded as endowed with special functions distinct from the less prominent portions of the cerebral surface, the sulci, which are continuous and identical in structure with them, but because they afford good indications of a particular arrangement of the surface of the hemispheres by which one brain may be conveniently compared with another, whether they belong to the same or to different groups of animals.

The folded arrangement of the surface of the hemispheres, dependent as it is upon the grey matter, is evidently destined to bring the central and deep-seated parts of the hemispheres into union with a large extent of vesicular surface.

That the disposition of the convolutions, like that of all other parts of animal bodies, follows a particular law, is well illustrated by comparing the brains of different groups of animals, in their gradation from the more simple to the more complex.

M. Leuret very justly makes a distinction between those convolutions which are constant, and to be found throughout the whole series of convoluted brains, occupying the same position, and differing only in size and extent of connections, and those which are not constant, even in the brains of the same group of animals, but are dependent on the extent of the primary ones, and the connections which they form with others near them. According to this idea we may classify the convolutions as *primary* and *secondary*.

The primary convolutions are all formed after one type. Of this, as M. Leuret suggests, the brain of the fox may be taken as the basis. The fissure of Sylvius is well marked in this brain; it is bounded by a prominent convolution, which encloses it above, below, and behind—thus forming a curve, the concavity of which is directed forwards and downwards. Above and behind this we find a second convolution forming a similar curve and parallel to the first. It exhibits a slight undulation, and is marked by a short fissure—signs of advancing complication. Still further back and upwards there is a third convolution, parallel and curved similarly to the second; this bifurcates at one point. Above all, near the summit of the hemisphere, a fourth is found disposed in the same curved manner, but exhibiting some sinuosities or undulations at its anterior portion. A fifth convolution exists on the inferior surface of the anterior lobe and rests upon the roof of the orbit. Leuret designates it *the supra-orbitar convolution*. The sixth convolution is of great extent; the principal portion of it is found on the inner surface of each hemisphere above the corpus callosum; in front it bends downwards and backwards to the fissure of Sylvius, and behind it extends to the middle lobe and forms the hippocampus major. This convolution exists in a high state of developement in the human brain, and has attracted very generally the attention of anatomists. Foville describes it by the name *convolution d'ourlet*.

Such is the most simple arrangement of the convolutions. The complication of this takes place by undulations being formed in the convolutions themselves, by a subdivision of them at certain situations, by the junction of neighbouring ones through smaller folds crossing the sulci between them, and in the highest classes by the addition of totally new convolutions.

Animals, whose brains have nearly the same degree of developement as that of the fox, have exactly the same convolutions, differing, however, somewhat in point of size. This increase of size is denoted by undulations formed in the course of convolutions throughout more or less of their extent. The dog may be taken as an example. M. Leuret states that, in comparing the brains of several dogs together, he found with all of them the same convolutions, differing only in the extent of undulations and the number of depressions, both of which were greatest in the largest brains. The brain of a large mastiff *(chien dogue)*, a good watch-dog, of such great ferocity that he attacked the person who fed him, had all the convolutions very large and much undulated, with numerous depressions in them.

A group of animals, consisting of the cats and the hyena, exhibits another stage of increase in the developement of convolutions. The same type prevails as in the fox and dog; four external convolutions, one internal, and a supra-orbitar. These convolutions, however, are united to each other at numerous points by means of small folds crossing the sulci. These uniting folds form the *secondary* or *supplementary* convolutions. Nearly all the primary convolutions have supplementary ones connected with them.

A group, which includes the sheep and other ruminant animals, exhibits much more complication in the cerebral convolutions, but still preserves the same type. The undulations and the supplementary convolutions are more numerous. The primary appear less numerous because less distinct. The anterior part of the internal convolution is much increased in developement, and the supra-orbitar is much more complex. In the fissure of Sylvius some small convolutions are found which are the first developement of those which in the human subject constitute the *insula* of Reil.

In the brain of the elephant new convolutions are added. These consist of folds passing in a perpendicular direction; the primitive convolutions always taking a longitudinal course. These latter are divided by the former into an anterior and a posterior set. Others are found above and in front of the fissure of Sylvius; three superior convolutions are found, the continuations of which backwards are situate above the internal convolution. All the convolutions of the elephant are remarkably undulating and exhibit numerous depressions. The brain of the whale is very similar to it in this respect, and both resemble that of man.

Monkeys have not the tortuous or complicated convolutions which are found in the whale and elephant. Yet the developement of the hemispheres at their posterior part, the general form of the brain, the extent and inclination of the fissure of Sylvius approximate the brain of monkeys to that of man much more nearly than the whale's or elephant's, which, notwithstanding their complicated convolutions, are generally inferior in organization, and resemble the brains of other Mammalia. The internal convolution in monkeys is simple ; below and behind it forms the hippocampus, from which convolutions are prolonged backwards, forming the posterior lobe. Two superior convolutions are met with above the fissure of Sylvius, between which is placed a very constant transverse fissure, called *the fissure of Rolando*. The orbitar convolutions are largely developed.

In comparing the human brain with that of the inferior animals, we notice great exactness of symmetry between the convolutions of opposite hemispheres in the latter, and the want of it in the former. It cannot, however, be said that the convolutions of opposite hemispheres in the human subject are absolutely unsymmetrical. A careful examination will show that the same convolutions exist on each side, but apparently of different sizes, and not corresponding closely as regards situation. My meaning will be more readily understood by referring to *fig.* 21, p. 138, where the same numbers have been affixed to corresponding convolutions. No. 1 on the right has a certain general resemblance with No. 1 on the left, which would be much more perfect but for the fissure which marks the convolution of the right hemisphere. Again, Nos. 2, on opposite sides, resemble each other so closely that their symmetrical relation cannot be doubted. The likeness, however, is impaired by slight fissures in the convolution on the left which do not exist in that on the right side. Nos. 3 and 3 evidently correspond, but that of the right side is the larger and more undulating. And it may here be remarked that this great developement of the convolution marked 3 on the right side affects materially the position, relations, and shape of those in its neighbourhood, by throwing them backwards or outwards and altering their form. Thus the position and shape of convolution 4 seems evidently modified by the large posterior undulations of convolution 3. In the brain from which the figure was taken, the convolutions on the right side are evidently larger and more highly developed than those of the left. It does not appear that there is any constancy with respect to the relative size of the convolutions of the right and left side, sometimes one side predominating, sometimes the other; nor have we any clue to discover the cause of the difference between the two hemispheres, or the reason of the variation as regards predominance of size.

In the imperfectly developed brains of the infant or young child, the convolutions are quite symmetrical. They are so likewise in idiots, or persons of very inferior intellect, and, as has been already stated, in some Negro brains.

The following convolutions of the human brain are constant in their position, although they differ much in different brains in size and developement.

1. *The internal convolution,* or that of the corpus callosum, called by Foville *convolution d'ourlet (processo cristato,* Rolando). The principal portion of this convolution is above and parallel to the corpus callosum: in front it curves down parallel to the anterior reflection of the corpus callosum, as far as the locus perforatus, connecting itself with some of the anterior convolutions. Behind it passes in a similar manner round the posterior reflexion, connecting itself with some of the posterior convolutions, and in the middle lobe forming the hippocampus major, the anterior extremity of which is situate

immediately behind the fissure of Sylvius and locus perforatus. Its horizontal portion appears to be connected with some nearly vertical ones, which seem indeed to branch off from it. (*Fig.* 35, O.)

This is the most constant and regular convolution of the brain. It exhibits with its fellow of the opposite side very exact symmetry. Its inferior or concave border is smooth and uninterrupted, and forms the superior boundary of a sulcus, which intervenes between it and the surface of the corpus callosum. It forms, to use Foville's expression, a hem or selvage to the cortical layer of the cerebral hemisphere. The fibrous matter which is inclosed by the cortical layer of this convolution consists of longitudinal fibres following the same general direction, a large number of them no doubt bending inwards into the cortical layer. These fibres are evidently commissural in their office, and will be referred to by-and-bye as constituting the *superior longitudinal commissure*.

The free margin of this convolution varies in its characters in different brains, according to the degree of tortuosity it exhibits, and the number of small fissures which are met with in it. The small folds which connect it with other convolutions on the inner surface of the hemisphere vary in number, and are generally found most numerous at its posterior part. Some of these folds are not distinctly visible unless the sulcus above it has been freely opened, as they are situated quite on its floor.

2. *The convolution of the Sylvian fissure.*—This convolution forms the immediate boundary of this great fissure. We have seen its early developement in the simple brain of the fox, and we may observe it gradually rising in complexity through all the intermediate stages up to the most highly developed brains. In the elephant it is remarkably tortuous, and is connected anteriorly as well as posteriorly with convolutions which pass to the anterior and superior and to the posterior part of the

THE CONVOLUTIONS.

Fig. 35.

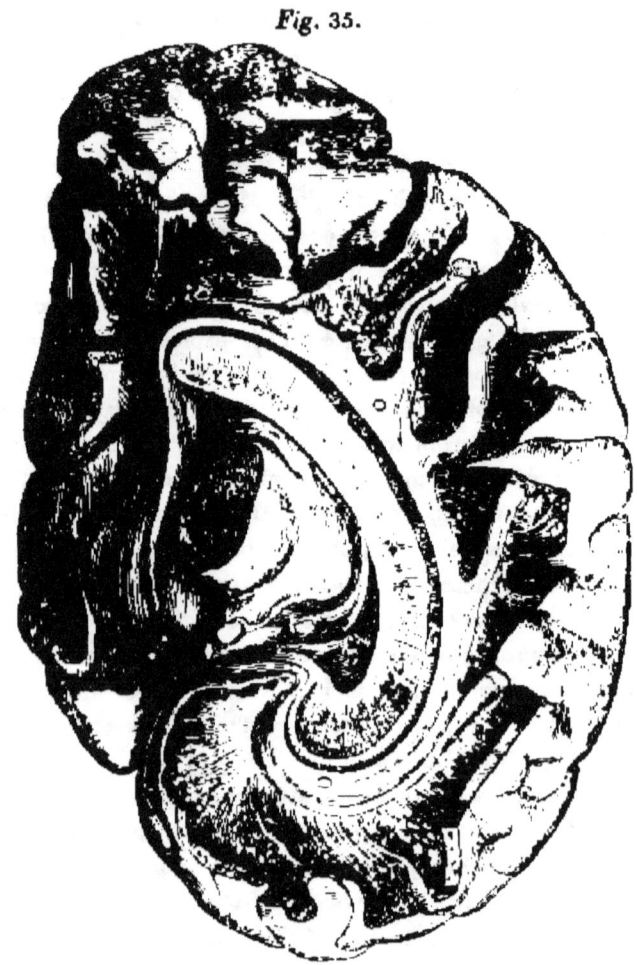

Internal surface of the left hemisphere of the brain, shewing the connections of the internal convolution and the band of longitudinal fibres by which it is formed (d'ourlet).

C, C, corpus callosum ; O, O, O, internal convolution ; *b*, septum lucidum ; *a*, anterior commissure ; *f*, fornix ; *c*, superior layer of the crus cerebri ; *d*, inferior layer of the same separated from the former by the locus niger.

The fibres of the internal convolution are seen in the middle lobe extending to the hippocampus major.

brain.* In man it is also very tortuous, and the numerous folds which pass from it forwards or backwards, forming primary or secondary convolutions, render it difficult to isolate it sufficiently for the anatomist to follow it throughout its entire course. Its inner border is likewise interrupted by the connections which it forms with the convolutions of the floor of the Sylvian fissure.

3. Within the fissure of Sylvius we find that remarkable group of convolutions called by Reil *insula,* the island. It consists of a series of small folds radiating from a common centre and connected with the convolution last described by still smaller folds, which cannot be seen unless when the fissure has been very freely laid open. The centre from which the convolutions radiate is the apex of a cone, the base of which adheres to the floor of the fissure.

4. On the inferior surface of the anterior lobe there is a pair of longitudinal convolutions which enclose between them the fissure of the olfactory process. The external of these convolutions is continuous with the convolution of the Sylvian fissure.

The numerous secondary convolutions which are found over the surfaces of the brain render it difficult to distinguish the primary ones. These latter are indicated by the antero-posterior course which they take—the former being more or less vertical. The largest and most tortuous convolutions are found on that part of the external surface which corresponds to the parietal bone. Next to them, in point of size, are the convolutions of the anterior lobe, but the smallest of all are those of the posterior lobes.

The hippocampi, major and minor, are constant convolu-

* See Leuret, pl. xiv. representing the external surface of the elephant's brain.

tions, which project into the lateral ventricles, the latter into its posterior, the former into its descending horn.

In general the constituent fibres of the white matter of the convolutions converge from the inner surface of the cortical layer to the centrum ovale, or if followed from the centrum ovale, they radiate to the grey surface, whether of a convolution or of a sulcus. A remarkable exception is in the case of the internal convolution, the fibrous matter of which constitutes, as has been already explained, a longitudinal commissure. The thickness of the cortical layer is pretty uniform, at least relatively to the size of the folds themselves. Throughout its entire thickness it is mixed with fibres, which are most numerous at its adherent surface, but extremely few and scattered at its free surface.

In hydrocephalus the convolutions disappear. The fibrous matter becomes greatly expanded by the fluid accumulated in the ventricles, and when its expansion has gone so far as to equal the grey surface, the folded character of the latter disappears. This takes place precisely in the same way that the rugæ of the contracted stomach (as before referred to) become obliterated when the muscular coat relaxes and allows the full distension of the organ.

Mayo supposes that other fibres are found in the convolutions besides those which are continued into the centrum ovale. These are commissural ones, which pass from convolution to convolution—either between adjacent or distant ones—forming arches the convexities of which are directed to the centrum ovale. I have never succeeded in satisfying myself of the existence of such fibres either in the fresh brain or in that preserved in spirit. If they exist, it is evident that they must be commissural between particular convolutions. The same anatomist supposes that similar commissural fibres connect the laminæ of the cerebellum.

The principal bulk of the hemispheres is formed by fibrous substance. This is shown by the horizontal section which displays the *centrum ovale*. These fibres radiate from those surfaces of the optic thalami and corpora striata which are in contact with the substance of the hemisphere. Most of the fibres which emerge from these gangliform bodies pass to the grey matter of the convolutions. Some, however, turn inwards towards the mesial plane, and contribute to form the corpus callosum by their union with those of the opposite side.

It cannot be supposed that all the remaining fibres of the hemispheres, after subtracting those which form the corpus callosum, pass through the thalami and corpora striata. The disproportion in bulk between the aggregate of these fibres and those of the medulla oblongata is too striking to admit of such an hypothesis. The most probable view is that the proper fibres of the hemispheres are implanted in the vesicular matter of each of these bodies, and do not pass beyond them; and that these bodies receive other fibres from below, which are continuous with those of the anterior pyramids and the olivary columns. But it must be regarded as doubtful whether any of the fibres of the hemispheres pass through the corpora striata and optic thalama, and become continuous with fibres of the medulla oblongata.

Corpora striata and optic thalami.—The corpora striata and optic thalami bear a strong resemblance in general character and structure to ganglia. They are ovoid masses placed between the fibrous substance of the hemisphere on the one hand, and the medulla oblongata on the other. These bodies, which are best displayed by laying open the lateral ventricles (p. 146), are very closely united to each other. The corpus striatum is placed a little in front and to the outside of the thalamus. It is pear-shaped: its thick end is directed forwards and inwards, and it gradually tapers backwards into a caudate process of considerable length, which winds downwards, forwards, and inwards into the descending cornu of the lateral ventricle, at the anterior extremity of which it terminates. Placed on the outside of the thalamus, it seems to embrace it there, and to adhere very intimately to it. The tænia semicircularis lies in a groove between the two bodies, and as it were constricts their connecting fibres.

The corpus striatum is of a dark grey colour. A considerable portion of it projects free into the cavity of the ventricle, forming an extensive convex surface there. The rest is firmly imbedded in the fibrous substance of the hemisphere, and in position corresponds to the base of the insula, which for that reason has been called *the lobule of the corpus striatum*. The free surface as contributing to form the ventricular wall is covered by the lining membrane of the ventricle and a layer of nucleus-like particles; it is traversed by several veins. This surface is limited on the outside by the plane of fibres, which, after emerging from it, incline inwards and contribute to form the corpus callosum. On the inside it is limited by the tænia semicircularis, which separates it from the optic thalamus. That portion of the free surface which is seen in the inferior horn of the ventricle has, as already stated, the appearance of a caudiform prolongation of the upper portion; this probably arises from the diminution of the body in thickness at its inferior part, the portion which belongs to the inferior cornu forming the apex of a cone, of which the upper convex portion forms the curvilinear base.

When sections are made through the deep portion of each corpus striatum, the surfaces appear to be traversed by very numerous bundles of fibres. It is necessary that the sections should be made obliquely from below upwards in a direction parallel to the inferior layer of the crus cerebri. The bundles of fibres are thicker and more closely approximated to each other inferiorly; but as they ascend, they diverge, and radiate, some forwards, others outwards, and others backwards; some pass nearly vertically upwards. A section made quite in the horizontal direction cuts all these fibres more or less transversely, so that the cut surface presents a grey colour interspersed with white spots of variable size, according as the bundles have been cut transversely or obliquely; but when the section is made obliquely, as above directed, then the surface presents a striated appearance like numerous and regular white veins in a dark marble, the bundles of fibres being cut lengthways.

In tracing the bundles of fibres through the corpus striatum, we find that they divide and subdivide and occasionally anastomose. Each subdivision becomes clothed as it were with grey matter, which fills up the space between it and the adjacent ones. The grey matter ensheathes these bundles of fibres, as the areolar tissue does the coarse fascicles of large muscles, and it may be dissected away from them, as we remove the areolar tissue from the muscular bundles.

It is an important problem to determine the exact source of these fibres and their precise destination. There can be no doubt that many of them are continuous with the inferior plane of the crus cerebri. Of these, the major part are usually supposed to pass through to the white substance of the hemisphere or of the corpus callosum; some, however, undoubtedly proceed no farther than the corpus striatum. The other fibres which are found in this body may be viewed as taking their point of departure from its vesicular matter, and radiating, some outwards into the centrum ovale, others backwards to the optic thalamus, forming a bond of connection with that body. It must be borne in mind that, as the corpus striatum is a body of considerable thickness, the fibres which emerge from it must proceed in very different planes and with varying degrees of obliquity. Other fibres are found in the corpora striata, which however do not contribute to its striation. These are the fibres of the anterior commissure.

From a comparison of the small amount of fibrous matter in the inferior plane of the crus cerebri with the immense mass which forms the white substance of the hemispheres, (even if we exclude those fibres which form commissures,) it is impossible, as already remarked, to suppose that the latter is derived from the former only; nor, indeed, can it be admitted that even the greater part of the fibrous matter of the hemispheres is continuous with that of the crura, whether on their superior or inferior plane. A considerable portion of them doubtless, when traced from the hemispheres downwards, will be found not to pass below the corpora striata or optic thalami.

We may regard the corpus striatum as a mass of grey matter with fibres implanted in it which connect it with the other parts of the encephalon. These parts are, 1st, the hemispheres; 2d, the optic thalami; 3d, the crura cerebri, mesocephale, and medulla oblongata. Of these last fibres it is possible, (but I am disposed to think far from certain,) that some of those which form the inferior layer of the crus pass through the corpora striata, and diverge among the other fibres of the centrum ovale.

Thus the corpora striata are connected to the optic thalami by fibres which pass from their concave or inner border to those bodies; to the convolutions of the brain by fibres continuous with some of those which form the white substance of the hemisphere, and we have seen that the convolutions of the *insula* have a very close relation to them; to the mesocephale and medulla oblongata by the fibres which form the inferior layer of the crus; and to each other by those which, emerging from them, contribute to form the corpus callosum, and also by the anterior commissure.

The vesicular matter of the corpora striata does not differ from that of the convolutions. It is traversed by a multitude of fibres. These, however, do not form any intricate interlacement as in ganglia, but are collected into bundles of very variable size; the largest being placed at the inferior part of the body, the smallest towards the hemispheres. The free or ventricular portion of the corpus striatum contains comparatively few fibres. When a portion of the striated part of the body is examined under the microscope, the nerve-fibres, of which the bundles are composed, appear to be reduced to their smallest size, and to be very compactly applied to each other, so that they transmit very little light, and therefore put on the appearance of dark cylinders. It is only by very high powers that we can discover their fibrous structure. In many of the bundles the fibres appear to terminate at one extremity as if by forming an adhesion around a large vesicle, faint indications of the nucleus and nucleolus of which may be sometimes seen through the surrounding fibres. The appearance which the fibrous cylinders which exhibit this structure present calls to mind very strongly the representation of the nucleus of a comet with its tail. And this peculiarity of structure may be adduced as an argument that many if not the greater number of the fibres of the striated body form intimate connections with the elements of its vesicular matter.

Optic thalamus.—In the internal concave surface of each corpus striatum, the optic thalamus is placed. The latter body is therefore posterior and internal to the former. The lighter colour of the optic thalamus distinguishes it at once from the corpus striatum. The optic thalami come into close relation to each other by their inner surfaces, which form the lateral boundaries of the third ventricle.

Each optic thalamus, like the corpus striatum, presents a free and an attached portion. The former projects into the ventricle—*the intra-ventricular* portion; the latter adheres to the inner side of the corpus striatum and to the mass of the hemisphere, and posteriorly, to the olivary columns, the quadrigeminal tubercles, and the processus cerebelli. The superior surface is free and forms part of the floor of the lateral ventricle; the internal surface is likewise free and forms the lateral wall of the third ventricle, being, however, interrupted in a very small space by the adhesion of the soft commissure. A portion of its external and posterior surface is also free, and projects backwards and outwards into the inferior horn of the lateral ventricle, presenting a pointed extremity in that situation. These free surfaces are smooth and moist, being covered by the membrane of the ventricles. The velum interpositum, which is itself overlapped by the fornix, rests upon the superior surface of the optic thalamus.

The optic thalami are placed obliquely, so that they are nearer each other at their anterior than at their posterior extremities. Each measures about an inch and a half in length, nine to ten lines in height, and about eight lines in breadth. In colour they are very much lighter than the striated bodies, and they appear to be covered with a delicate layer of fibrous matter. A band of fibrous matter passes along the inner surface of each from behind forwards, which posteriorly is connected to the pineal gland, and forms, with its fellow, the peduncles of that body.

Beneath the posterior free extremity of the thalamus, situated in the angle between that body and the superior surface of the crus, we find a small rounded eminence of a darkish grey colour perforated by very numerous foramina for the transmission of bloodvessels. This is the *corpus geniculatum internum*. Lower down and more external and anterior, there is another similar body, somewhat smaller and less dark, the *corpus geniculatum externum*. Both of these bodies are connected with the quadrigeminal tubercles. A band of fibrous matter passes from the testes to the external geniculate body, and from the nates to the internal one.

In point of structure the thalamus resembles a ganglion much more closely than does the corpus striatum. A light reddish grey is the colour of the surface when cut into; it has been not inappropriately compared to that of coffee mixed with a large proportion of milk (café au lait). When thin sections are examined, they are found to consist of very numerous fibres interlacing freely, with nerve vesicles occupying their intervals. The fibres are not collected into bundles as in the corpus striatum, nor do they take a radiating course in the thalamus. The reticulation which they form is very like that in the ganglia on the posterior spinal roots. The vesicular matter appears to be chiefly, if not entirely, of globular vesicles.

The fibres of each optic thalamus are extremely numerous, and have most extensive connections. Along its ventricular surface they are evidently continuous with those of the hemisphere, which appear to radiate from it to the grey matter of the convolutions. Posteriorly the fibres of the processus cerebelli ad testes and those of the olivary columns pass into it. The anterior pillars of the fornix are connected with it in front, and derive from it some nervous fibres; and below and within, a cylinder of fibres emerge from it to the mammillary bodies. Thus the optic thalami are connected with the hemispheres on one hand, with the olivary columns and with the cerebellum on the other hand. The quadrigeminal tubercles placed upon the processus cerebelli may have some connection with them through the olivary columns. Although these bodies have been viewed as having a special connection with the optic nerves, it does not appear that these nerves have any relation to them but through the geniculate bodies or the quadrigeminal tubercles. It is important to bear in mind respecting the optic thalami that they are directly continuous with the superior portion of the crus cerebri, so that in viewing a vertical section of the encephalon we see no line of demarcation between. The thalamus grows, as it were, from the superior extremity of the crus; it is recognised from the latter by its swelling into an ovoidal mass. It is emphatically, as Willis long ago expressed it, an *epiphysis* upon the crus cerebri; and in this sense it may be classed with the striated bodies and the quadrigeminal tubercles, which, with the thalami, form a series of gangliform masses, disposed in pairs, one beyond the other.

The geniculate bodies, although intimately connected with the optic thalami, appear to be distinct from them, but very similar in structure. A section made into the thalamus through either of them shows a distinct line of demarcation between them. The optic tracts adhere to the lower surface of each thalamus by their inner margins, and when followed backwards are found to form a very evident connection with both the geniculate bodies.

Corpora mammillaria.—These bodies may be conveniently noticed here, as forming part of the series of gangliform masses in connection with the brain. They are of a spherical shape, covered on the exterior by very pure white matter, which is apparently derived from the anterior pillars of the fornix. When cut into, they are found to consist of a mixture of vesicular and fibrous matter, surrounded by a thin cortex of the latter. Microscopic examination proves the structure of the interior substance to be of the same nature as that of ganglia, and to resemble the optic thalamus.

The fibrous matter is connected, at the upper part of each body with the anterior pillar of the fornix, and on the outside with a fasciculus of fibres from the optic thalamus. It has been supposed that these two bundles are continuous, and that the mammillary body results from a twisting of the anterior pillar of the fornix as it changes its direction to pass into the substance of the thalamus. But the ganglionic structure of the mammillary bodies is unfavourable to this view, and renders it more probable that they are independent structures, exercising proper functions as nervous centres; and the constancy of these bodies in, at least, the mammiferous series, increases this probability.

Of the commissures of the brain.—A large number of fibres connected with the hemispheres or the two gangliform bodies just described, seem to serve the purpose of connecting different parts, either on the same side of the mesial plane, or on opposite sides of it. Those on the same side constitute the longitudinal commissures; those on opposite sides the transverse.

Of the longitudinal commissures.—These are four in number. They all evidently belong to the same system of fibres, separated from each other by the developement of intermediate parts. That which is on the highest plane is the *superior longitudinal commissure,* which is the fibrous matter of the internal convolution. Internal and a little inferior to this is a second, very small, band of fibres, *longitudinal tract,* (Vicq. d'Azyr,) which passes from before backwards along the middle of the corpus callosum, parallel to the superior longitudinal commissure, from which it is separated only by the grey matter of the convolution. Both these commissures are separated by the corpus callosum from a third, which takes a course parallel to them, namely, the *fornix,* which occupies a plane considerably inferior to both. External to this and separated from it by the ventricular projection of the optic thalamus, we find a fourth band, which passes parallel to the fornix: this is the *tænia semicircularis.*

As these parts have been already described, it will be unnecessary to do much more at present than indicate the connections which they serve to maintain.

1. *The superior longitudinal commissure* connects the convolutions of the inferior surface of the anterior lobe with the hippocampus major; and as its fibres pass above the corpus callosum they form connections with some of the other convolutions on the internal surface of the hemisphere (*fig.* 35).

2. *The longitudinal tracts* of the corpus callosum may be traced from about the same region of the inferior surface of the anterior lobe as the preceding commissure, near the perforated space, and they pass backwards, winding over the posterior reflection of the corpus callosum to its inferior surface.

3. The *fornix** is, next to the corpus callosum, the most extensive of the cerebral commissures. That it consists of longitudinal fibres cannot be doubted. Although commonly described as a single structure united at the body of the fornix, and spreading backwards and forwards by its crura, it nevertheless is distinctly divisible along the middle line into two perfectly symmetrical portions. The adhesion of the transverse fibres of the corpus callosum on its upper surface, and of the terminal fibres of its posterior reflection on its inferior surface which form *the lyra,* is the principal bond of union of these two lateral halves of the fornix.

The separation of its anterior pillars in front affords strong indication of its double form. These pillars pass downwards in a curved course, through the grey matter of the tuber cinereum, to the mammillary bodies, which are connected to the optic thalami by a bundle of fibres which may be easily traced into them. This bundle is described by Reil as the *root* of the anterior pillar of the fornix. From the gangliform structure of the corpus mammillare, I prefer to regard this band as a

* This commissure is called *voûte à trois piliers* by the French, *Trigone cérébrale,* Chaussier: Ψαλις, σωμα Ψαλιοειδης of the Greeks. Mr. Solly has given an excellent delineation of the fornix in his work on the brain,. pl. ix.

medium of connection with the optic thalamus, and to trace the anterior pillars to the mammillary body.

The parts with which the fornix is connected in front are the optic thalami, the mammillary bodies, and the septum lucidum, which consists of fibres having the same physiological import as those of the corpus callosum, altered however in direction by the backward position of the anterior pillars, which adhere to the body of the fornix.* The tuber cinereum and the grey matter which adheres to the lower half of the inner surface of each optic thalamus, are connected with it; and as its anterior pillars pass upwards in this situation, they receive fibres from the neighbouring convolutions. These pillars remain separate from the mammillary bodies to the foramen of Monro, where they adhere to each other and form the apex of the body of the fornix. Traced backwards, the fibres of the fornix pass into the posterior and inferior horns of the lateral ventricle. In the former they connect themselves with the hippocampus minor by expanding over it, and in the latter they spread over the hippocampus major, forming the posterior pillar of the fornix, or tænia hippocampi.

The relation of the anterior pillars of the fornix to the foramen commune anterius has been already sufficiently described. Superiorly the fornix adheres to the corpus callosum or to the septum lucidum, and the anterior commissure crosses in front of its anterior pillars, and almost touches them.

4. The fourth longitudinal commissure is the *tænia semicircularis*. It may be traced from the corpus mammillare outwards and backwards in the groove between the optic thalamus and corpus striatum into the inferior horn of the lateral ventricle, where its fibres mingle with those of the middle lobe. It is evidently part of the same system of fibres with the fornix.

* Mr. Solly, however, regards the septum lucidum as being formed of longitudinal fibres, and therefore more analogous to the fornix than to the corpus callosum.

Take away the corpus callosum, the grey matter of the internal convolution, the ventricular prominence of the optic thalamus, and all these commissures fall together and become united as one and the same series of longitudinal fibres.

It is very remarkable how few fibres pass between the great mass of the cerebrum and the cerebellum. The *processus cerebelli ad testes* are the only fibres which can be regarded as forming commissures between these two segments of the encephalon, and they are of the nature of longitudinal commissures.

The *transverse commissures* are the *corpus callosum*, the *anterior commissure*, the *posterior commissure*, the *soft commissure*.

1. The *corpus callosum*, so called according to some from the density of its tissue, is the great commissure of the lateral halves of the brain proper—*commissura cerebri maxima* of Soemmering. The fasciculated character of this structure is as obvious as that of any nerve in the body, and the direction of its fibres is clearly from one hemisphere to the other.

From the description already given of the corpus callosum, it is evident that its fibres sink into the white substance of each hemisphere above the level of the corpora striata, as well as into that of the anterior and of the posterior lobes. By its principal or horizontal portion it connects the white matter of the lesser centrum ovale of each side, and by the fibres which form the anterior and posterior reflexions it connects the anterior and posterior lobes. It needs only a very superficial dissection to ascertain thus much.

To determine the precise fibres with which those of the corpus callosum are continuous, and the relation which they bear to the lateral ventricles, demands a much more minute dissection. This must be done, according to the directions of Foville, who has given a most elaborate description of this commissure, by carefully separating it in the transverse direction from the internal convolution, on a hardened brain. Pursuing the dissection in this direction, it may be detached from the substance of the hemispheres as far outwards as the external border of the corpus callosum and optic thalamus. Along this edge the fibres curve downwards and inwards, and appear to become continuous with some of those which radiate from those bodies. The anterior and posterior fibres enclose the anterior and posterior horns of the lateral ventricles in radiating forwards and backwards from the corpora striata and optic thalami to those parts of these cavities.

This view of the connections of the corpus callosum would indicate it to be a commissure between the thalami and corpora striata, or between the crura cerebri, as Tiedemann supposed, rather than between the hemispheres. Nothing is more difficult than the dissection of the fibres of the corpus callosum beyond the internal convolution: and it cannot be regarded as in any degree certain that the connections of its fibres are limited to those above described.

The developement of the corpus callosum in the fœtus, prior to that of the hemispheres or convolutions, is favourable to the view of its connections maintained by Tiedemann and Foville. Comparative anatomy is, however, more in accordance with the opinions of Gall and Reil, that it is a commissure between the convolutions of opposite sides. It exists only in those animals in which convolutions are amply developed. In Fish, Reptiles, and Birds it is absent, and the Mammalia with least perfect brains, as the Rodents and Marsupialia, have it either imperfectly developed, or not at all.

The corpus callosum is a stratum of considerable thickness. Its fibres are situate on different planes, which interlace with each other so much as to render it impossible to separate a single layer for any distance, and the difficulty is much increased as the fibres approach the white mass of the hemispheres.

2. *The anterior commissure* may be regarded as truly a bond

of connection between the hemispheres, as well as between the corpora striata. It is a cylindrical cord of fibrous matter, very definite in its course and connections, and easily traced throughout its entire extent. Its situation in front of the anterior pillars of the fornix has been already described. If followed on either side from this central portion, it may be traced through the grey matter of the anterior and inferior portion of each corpus striatum into the fibrous matter of each middle lobe of the brain. Its course is curved with convexity directed forwards. As it passes outwards on each side it becomes flattened, and after it has traversed each corpus striatum, it expands considerably and its fibres radiate extensively. It may be stated to connect the convolutions of the middle lobes and the corpora striata.

3. *The posterior commissure* is a band of fibres, extended between the posterior extremities of the optic thalami, upon which rests the base of the pineal body. Those fibres, which immediately support that body, have been distinguished as the *pineal commissure;* but as they are evidently part of the same system as those which constitute the posterior commissure, there seems no good reason for separating them.

4. *The soft commissure* is also extended between the thalami. It is composed of vesicular matter with fibres, which pass from one side to the other. The intermixture of its fibres with vesicular matter distinguishes it from the other transverse commissures already described. A layer of a similar nature connects the locus niger of each crus cerebri, and fills up the space between the crura—*interpeduncular space.* This has been already described as the *pons Tarini, posterior perforated space.* It consists of fibrous matter intermixed with vesicular, extended between the crura cerebri. It seems analogous to the soft commissure, and therefore entitled to be regarded as a commissure.

Of the manner in which the commissures connect the various parts between which they are placed, it is difficult to form an exact opinion. It is most probable that they form an intimate union with the grey matter of the segments which they serve to connect. It might also be conjectured that they are continuous with some of the fibres of the segments which they unite, or that they interlace with them in some intricate way, so as to come into intimate or frequent contact with them.

Tuber cinereum.—At the base of the brain we have already described a layer of pale grey matter which fills up the interval between the mammillary bodies and the optic commissure. It extends above the optic commissure forwards to the anterior reflexion of the corpus callosum, and forms intimate connections with the anterior pillars of the fornix, the optic tracts, the septum lucidum, and at the floor of the third ventricle with the optic thalami. It consists of vesicular matter with fibres, and resembles very much the soft commissure, to which it is very probably analogous in office.

The process called *infundibulum* or *pituitary process* extends from the inferior surface of the tuber cinereum down to the pituitary body. It is hollow, wide above, where it communicates with the third ventricle, and narrow below at the pituitary body. When cut across, fluid will escape from the third ventricle through it, and a probe passes readily from that cavity into it. It is composed of a layer of granules, derived, no doubt, from the epithelial lining of the third ventricle, and some vesicular matter with bloodvessels and fibrous tissue, which latter is derived from the pia mater, and the special sheath of arachnoid reflected upon it.

Pituitary body.—The process just described is the connecting link between the brain and that glandiform body, *the pituitary gland* or *hypophysis*. This body, situate in the sella Turcica, is of a rounded form, longer in the transverse than the antero-posterior direction, concave on its superior surface, into which the pituitary process is inserted. It is surrounded by dura mater, which projects over it, leaving an opening for the passage of the infundibulum.

The pituitary body is about six lines in its transverse diameter and three lines from before backwards: its weight, including the infundibulum, is about eight grains. It consists of two lobes, one anterior, the other posterior. The former is kidney-shaped and lodges the latter in the notch of its posterior edge. In point of size the anterior lobe is nearly double the posterior.

The colour of the posterior lobe is lighter than that of the anterior, and resembles that of the grey matter of the brain.

This body is proportionally larger in early life than at the later periods, and it is certainly more developed in the lower mammalia than in man. It is very large in fishes, and probably reaches its maximum of size in that class of animals.

The structure of the pituitary body resembles very much the grey cerebral matter. It is composed of large nucleated vesicles, surrounded by a granular matrix, with bundles of white fibrous tissue. This fibrous tissue either forms an essential element in its constitution, or accompanies the bloodvessels which are found in it in great numbers. Its substance is soft, but not so soft as the cerebral matter, and when pressed between the fingers is reduced to a greyish pulp, like the substance of an absorbent gland in an early stage of suppuration.

Earthy concretions have been occasionally but very rarely found in the pituitary body. This circumstance, its colour, its glandiform character, and its extra-cerebral situation in connection with the third ventricle, give it a certain degree of analogy to the pineal body. But in this latter nervous fibres have been found, of which I have failed to discover any trace in the pituitary, nor is the pituitary body connected to the brain by fasciculi of fibres as the pineal body is. The use of both is equally involved in obscurity; but from their constancy it may be argued that their function is not unimportant. It has been supposed that the pituitary body is a large ganglion belonging to the sympathetic system: this opinion, however, wants the all-important foundation of anatomy to rest upon, inasmuch as we find that the body in question is devoid of the anatomical characters of a ganglion. It may with more propriety be classed with the glands without efferent ducts; and from its numerous vessels, and its close relation to part of the venous system within the cranium, it may be connected with the process of absorption or removal of the effete particles of the brain.

Of the ventricles of the brain.—The third ventsicle results from the apposition of the lateral halves of the brain along the median plane, and the lateral ones from the folding inwards, above and below, of the convoluted surface of each hemisphere. They must not therefore be regarded as cavities hollowed in the substance of the brain: on the contrary, their walls must be viewed as part of the cerebral surface, and the eminences which project from them as convolutions. The corpora striata and optic thalami are from their structure entitled to be considered in this light, and still more the hippocampi, which, however, are somewhat complicated by the addition of the layers of white matter derived from the fornix.

The distinction between the lateral and the middle or third ventricle results from the developement of the corpus callosum and of the fornix, which form horizontal strata by which the ventricles are closed in above; and the extension of the anterior pillars of the fornix downwards, and the close application of the free margin of the body of the fornix to the optic thalami, assign more complete limits to the third ventricle.

The fourth ventricle is also evidently formed by the lateral adaptation of the symmetrical halves of the medulla oblongata. The iter is obviously a continuation of it closed behind by the quadrigeminal bodies and their connecting fibres. This ventricle remains open in the embryo, uncovered by any portion of the encephalon until the full developement of the cerebellum causes it to extend over it.

The fifth ventricle must be viewed as originally part of the third, which has been closed off by the full developement of the septum lucidum and fornix, and the union of their lateral halves along the median plane.

All these cavities are lined by a delicate membrane nearly allied to, if not identical with, serous membranes. It is covered by an epithelium, ciliated according to Purkinje and Valentin, beneath which are delicate fibres of areolar tissue exactly of the same kind as those found in connection with serous membranes. This membrane is reflected around the processes of pia mater which are found in the ventricles, and in this respect presents an additional point of analogy to the serous membranes, the portion which lines the walls of the ventricles corresponding to the parietal layer, and that which adheres to the pia mater resembling the visceral layer of those membranes. It is the reflection of this membrane from the walls to the enclosed pia mater which serves to shut off the ventricular cavity from the sub-arachnoid space, at the anterior part of the horizontal fissure, and at the inferior extremity of the fourth ventricle. If any communication take place between the intraventricular and sub-arachnoid fluid, it must be, as already remarked, by transudation through this membrane.

CHAPTER IX.

Of the circulation in the brain.

Haller calculates that the human brain receives rather more than one-fifth of the whole blood of the body. Whether this calculation be correct or no, it is certain that an organ of such great size, of such high vital endowments, so active, and which exerts so considerable an influence upon all other parts of the body, must necessarily require a large supply of the vital fluid. Four large arteries carry blood to the brain, namely, the two *internal carotids* and the two *vertebrals*. Each carotid penetrates the cranium at the foramen on the side of the sella Turcica, and almost immediately divides into three branches, the *anterior* and the *middle cerebral* arteries and the *posterior communicating* artery.

The *anterior cerebral arteries* supply the inner sides of the anterior lobes of the brain: they ascend through the great longitudinal fissure, and pass along the upper surface of the corpus callosum, giving off branches to the inner convolutions of both hemispheres of the brain. These arteries anastomose with each other just beneath the anterior margin of the corpus callosum by a transverse branch, called *anterior communicating artery*. The *middle cerebral arteries*, the largest branches of the carotid, pass outwards in the fissures of Sylvius, and supply the outer convolutions of the anterior lobes, and the principal portion of the middle lobes. At the inner extremity of each fissure of Sylvius numerous small branches of these arteries penetrate, to be distributed to the corpus striatum. The choroid arteries which supply the choroid plexus sometimes arise from these arteries, but also occasionally come from the carotid itself. The *posterior communicating artery* is an anastomotic vessel, which passes backwards along the inner margin of the middle lobe at the base of the brain, and communicates with the posterior cerebral artery, a branch of the basilar.

The *vertebral* arteries, having passed through the canals in the transverse processes of the cervical vertebræ, enter the cranium through the occipital foramen towards its anterior part. In their ascent they incline towards each other in front of the medulla oblongata, and at the posterior margin of the pons they coalesce to form a single vessel, the *basilar*, which extends the whole length of the pons.

The *vertebral* arteries furnish the *anterior* and *posterior spinal* arteries, and the *inferior cerebellar* arteries. These last vessels arise from the vertebrals very near their coalescence; they pass round the medulla oblongata to reach the inferior surface of the cerebellum, to which they are principally distributed.

From the basilar artery numerous small vessels penetrate the pons. At its anterior extremity it divides into four arteries, two on each side. These are, the two *superior cerebellar*, and the two *posterior cerebral* arteries.

The superior cerebellar arteries pass backwards round the crus cerebri, parallel to the fourth nerve, and divide into numerous branches on the upper surface of the cerebellum, some of which anastomose with branches of the inferior cerebellar artery over the posterior margin of the cerebellum. Some branches of these arteries are distributed to the velum interpositum.

The posterior cerebral arteries are the largest branches of the basilar. They diverge and pass upwards and backwards round the crus cerebri, and reach the inferior surface of the posterior lobe, anastomosing in the median fissure with ramifications of the anterior cerebral, and on the outside with branches of the middle cerebral arteries. Numerous small vessels pass from these arteries at their origin, and penetrate the interpeduncular space, and one or two are distributed to the velum. Shortly after its origin each of these arteries receives the posterior communicating branch from the carotid.

A remarkable freedom of anastomosis exists between the arteries of the brain. This takes place not only between the smaller ramifications, but likewise between the primary trunks. The former is evident all over the surface of the cerebrum and cerebellum. The latter constitutes the well-known *circle of Willis*. This anastomosis encloses a space, somewhat of an oval figure, within which are found the optic nerves, the tuber cinereum, the infundibulum, the corpora mamillaria, and the interpeduncular space or pons Tarini. The anterior communicating artery, between the anterior cerebral arteries, completes the circle in front. The lateral portion of the circle is formed by the posterior communicating artery, and it is completed behind by the bifurcation of the basilar into the two posterior cerebral arteries. Thus, a stoppage in either carotid, or in either vertebral, would speedily be remedied. The coalescence of the vertebrals to form the basilar affords considerable security to the brain against an impediment in one vertebral; and, should the basilar be the seat of obstacle, the anastomoses of the inferior cerebellar arteries with the superior ones would ensure a sufficient supply of blood to that organ. If either or both carotids be stopped up, the posterior communicating arteries will supply a considerable quantity of blood to the intracranial portions of them; or, if one carotid be interrupted, the anterior communicating branch will be called into requisition to supply blood from the opposite side.

Interruption to the circulation in both carotids and both vertebrals is productive of a complete cessation of cerebral action, and death immediately ensues, unless the circulation can be quickly restored. This was proved clearly by Sir A. Cooper's experiments on rabbits. The circulation may, however, be interrupted in both carotids, or in both vertebrals, without permanent bad effect; or in one carotid or one vertebral, provided the condition of the remaining vessels be such as not to impede the circulation in them. In cases where the neighbouring anastomotic branches are not sufficient to restore the circulation to a part from which it has been cut off by the obliteration of its proper vessel, the cerebral substance of that region is apt to experience a peculiar form of softening* or wasting, which is distinguished by the absence of any discoloration by the effusion of blood, or of any new matter.

The four great channels of sanguineous supply to the brain are continued up straight from the aorta itself, or from an early stage of the subclavian. The columns of blood contained in them are propelled very directly towards the base of the brain, through wide canals. Were such columns to strike directly upon the base of the brain, there can be no doubt it would suffer materially. Considerable protection, however, is afforded to the brain; first, by the blood ascending against gravity, during at least a great portion of life; secondly, by a tortuous arrangement of both carotids and vertebrals before they enter the cranial cavity; the carotid being curved like the letter S in and above the carotid canal, and the vertebral being slightly bent between the atlas and axis, then taking a horizontal sweep above the atlas, and after it has pierced the occipito-atlantal ligament, inclining obliquely upwards and inwards; thirdly, by the breaking up of the carotids into three branches;

---

* In the twenty-seventh volume of the Med. Chir. Trans. I have related a remarkable case in which white softening of one hemisphere followed the plugging of the common carotid on the same side by coagulum.

by the inclined position of the vertebrals, and by their junction into a single vessel, which takes a course obliquely upwards, and afterwards subdivides into smaller branches. Such arrangements most effectually break the force of the two columns, and, as it were, scatter it in different directions.

A further conservative provision is found in the manner in which the bloodvessels penetrate the brain. The larger arterial branches run in sulci between convolutions, or at the base of the brain; smaller branches come off from them, and ramify on the pia mater, breaking up into extremely fine terminal arteries, which penetrate the brain; or these latter vessels spring directly from the larger branches, and enter the cerebral substance. As a general rule, no vessel penetrates the cortical layer of the brain, which, in point of size, is more than two removes from the capillaries; and, whenever any vessel of greater size does pierce the cerebral substance, it is at a place where the fibrous matter is external, and that part is perforated by foramina for the transmission of the vessels. Such places are the locus perforatus, the interpeduncular space, &c. The capillaries of the cerebral substance are easily seen to possess an independent diaphanous wall, with cell-nuclei disposed at intervals. The smaller arteries and veins can also be admirably studied in the pia mater of the brain.

The venous blood is collected into small veins, which are formed in the pia mater at various parts of the surface, and in the interior of the brain. The superficial veins open by short trunks into veins of the dura mater, or into the neighbouring sinuses; the superior longitudinal, the lateral, and the straight sinuses receiving the greatest number. Those from the interior form two trunks, *venæ magnæ Galeni*, which pass out from the ventricles between the layers of the velum interpositum. The cerebral veins are devoid of valves.

We remark here, that the venous blood of the brain is returned to the centre of the circulation through the same channel as that of the dura mater, of the cranial bones, and of the eyeball: the internal jugular veins are the channel towards which the venous blood of the cranium tends. An obstacle, therefore, in both or either of these vessels must affect the entire venous system of the brain, or at least that of the corresponding hemisphere. A ligature tied tightly round the neck impedes the circulation, and may cause congestion of the brain. The bodies of criminals who have died by hanging exhibit great venous congestion, both of the walls and the contents of the cranium, in consequence of the strong compression to which the veins have been submitted.

We have seen that, when the blood of one carotid artery is cut off, the parts usually supplied by it are apt to become exsanguous and softened; and this is more especially the case if the vertebral be stopped up, or the circulation in it impeded. And it has been remarked, that these effects will follow the application of a ligature to either common carotid artery.

Notwithstanding these facts, a doctrine has received very general assent, and the support of men of high reputation, which affirms that the absolute quantity of blood in the brain cannot vary, because that organ is incompressible, and is enclosed in a spheroidal case of bone, by which it is completely exempted from the pressure of the atmosphere.

The cranium, however, although spheroidal, is not a perfectly solid case, but is perforated by very numerous foramina, both external and internal, by which large venous canals in the diploe of the bones communicate with the circulation of the integuments of the head as well as with that of the brain; so that the one cannot be materially affected without the other suffering likewise. And as the circulation in the integuments is not removed from atmospheric pressure, neither can that which is so closely connected and continuous with it be said to be free from the same influence. Still it must be admitted, that the deep position of the central vessels, and the complicated series of channels through which they communicate with the superficial ones, protect them in some degree from the pressure of the air, and render them less amenable to its influence than the vascular system of the surface.

If it were essential to the integrity of the brain that the fluid in its bloodvessels should be protected from atmospheric pressure (as the advocates of this doctrine would have us to believe), a breach in the cranial wall would necessarily lead to the most injurious consequences; yet, how frequently has the surgeon removed a large piece of the cranium by the trephine without any untoward result! Some years ago I watched for several weeks a case in which nearly the whole of the upper part of the cranium had been removed by a process of necrosis, exposing a very large surface to the immediate pressure of the atmosphere; yet in this case no disturbance of the cerebral circulation existed. In the large and open fontanelles of infants we have a state analogous to that which art or disease produces in the adult: yet the vast majority of infants are free from cerebral disease for the whole period during which their crania remain incomplete; and in infinitely the greatest number of cases in which children suffer from cerebral disease, the primary source of irritation is in some distant organ, and not in the brain itself.

It cannot be said that the brain is incompressible. That only is incompressible, the particles of which will not admit of being more closely packed together under the influence of pressure. That the brain is not a substance of this kind is proved by the fact that, while it is always undergoing a certain degree of pressure as essential to the integrity of its functions, a slight increase of that pressure is sufficient to produce such an amount of physical change in it as at once to interfere with its healthy action. Too much blood distributed among its elements, and too much serum effused upon its surface, are equally capable of producing such an effect.

Majendie's experiments, described at a preceding page, show that the brain and spinal cord are surrounded by fluid, the pressure of which must antagonise that which is exerted through the bloodvessels. The removal of this fluid disturbs the functions of these centres, apparently by allowing the vessels to become too full. The pressure exerted by the former may be called the fluid pressure from without the brain; that by the blood, the pressure from within. As long as these two are balanced, the brain enjoys a healthy state of function, supposing its texture to be normal. If either prevail, more or less of disturbance will ensue. Their relative quantities, if not in just proportion, will bear an inverse ratio to each other. If there be much blood, the surrounding fluid will be totally, or in a great measure, deficient; if the brain be anæmic, the quantity of surrounding fluid will be large.

The existence of these two antagonizing forces may be taken as an indication that either of them may prevail; and, therefore, from the presence of the cerebro-spinal fluid, we may infer that the actual quantity of circulating blood in the brain is liable to variation.

The cerebro-spinal fluid is a valuable regulator of vascular fullness within the cranium, and a protector of the brain against too much pressure from within. So long as it exists in normal quantity it resists the entrance of more than a certain proportion of blood into the vessels. Under the influence of an unusual force of the heart an undue quantity of blood may be forced into the brain, the effects of which will be, first, the displacement of a part or of the whole surrounding fluid; and, secondly, the compression of the brain.

When the brain receives too little blood, the requisite degree of pressure will be maintained, and the healthy cerebral action preserved, if the surrounding fluid do not increase too rapidly. But if the brain be deprived of its due proportion of blood by some sudden depression of the heart's power, there is neither time nor source for the pouring out of new fluid, and a state of syncope or of delirium will ensue. Such seems to be the explanation of those cases of delirium which succeed to hæmorrhages, large bleedings, or the sudden lighting up of inflammation in the pericardium or within the heart, or in some other organ of great importance to life. In nearly all these cases, however, it is important to notice that the blood is more or less damaged in quality, deficient in some of its staminal principles, or charged with some morbid matter; and this vitiated state of the vital fluid has no doubt a considerable share in the production of the morbid phenomena.*

* The subject of the circulation in the cranium has been very ably discussed by Dr. G. Burrows in the Lumleian Lectures for 1843, Lond. Med. Gazette, vol. xxxii.

CHAPTER X.

Origins of the encephalic nerves.

There are no common characters possessed by these nerves, such as have been enumerated at a preceding page for the spinal nerves. They are, however, disposed in pairs, and are quite symmetrical. With the exception of the olfactory, optic, and third pair, they are all connected with the mesocephale or medulla oblongata.

The arrangement of these nerves originally proposed by Willis has been so long adopted in this country and on the continent that no advantage would arise from abandoning it, unless some other of an unexceptionable nature could be substituted for it.

Twelve pairs of nerves are found in connection with the base of the encephalon. Five pairs have been so classed by Willis as to form two in his arrangement, three pairs being allotted to his eighth pair of nerves, and two to his seventh. Willis's arrangement, therefore, comprises nine pairs of nerves, which he enumerates, beginning at the anterior and passing to the posterior part of the base of the brain. These are the first pair or *olfactory* nerves; the second pair or *optic;* the third pair, *motores oculorum;* the fourth pair, *pathetici;* the fifth pair; the sixth pair, *abducentes oculi;* the seventh pair, including the *portio mollis* or *auditory* nerve, and the *portio dura* or *facial* nerve; the eighth pair, including the *glosso-pharyngeal,* the *pneumo-gastric,* and the *spinal accessory;* the ninth pair or *hypo-glossal.* The first cervical nerve or the *sub-occi-*

pital was considered by Willis as an encephalic nerve and counted as the *tenth pair*.

As the cranium may be shewn to be composed of the elements of three vertebræ, it has been attempted to prove that among these nerves some may be classed with the vertebral or spinal nerves. The fifth is obviously of this kind from its anatomical characters, namely, two roots; one, small, ganglionless; the other large, ganglionic; but with the former, which is analogous to the anterior root of a spinal nerve, the third, fourth, and sixth nerves may be conjoined from their similarity in structure and distribution. Thus one *cranio-vertebral* nerve is formed, the anterior root of which consists of the small portion of the fifth, the third, fourth, and sixth nerves; and the posterior or sensitive root, of the large portion of the fifth. A second cranio-vertebral nerve consists of the eighth pair, to which might be added the facial contributing to its motor portion; and a third is formed by the hypoglossal. The analogy, especially in the latter case, is far from being very obvious.

I shall occupy the remainder of this chapter with a brief account of the origins of the encephalic nerves.

I. *The olfactory nerves.*—This title has been commonly applied to two processes of cerebral matter which occupy a pair of deep fissures on the inferior surface of the anterior lobes of the brain. It is high time that this application of it should be discarded, and that these processes should be described under their proper names, *olfactory lobes*.

The olfactory lobes are two in number: one on the inferior surface of each anterior lobe of the brain. Each corresponds to the inferior margin of a fissure which passes upwards and outwards to a considerable depth into the substance of the brain, and is lined by pia mater. A straight convolution bounds each fissure on its inner side, separating it from the anterior extremity of the great longitudinal fissure.

Each olfactory lobe consists of a prismatic process of cerebral substance compounded of fibrous and vesicular matter, which terminates in front in an oblong bulb, composed chiefly of vesicular matter, and resting upon the cribriform plate of the ethmoid bone beside the crista galli. The upper sharp margin of the process in its two posterior thirds consists of vesicular matter; the lower portion of fibrous matter; the bulb is vesicular on its exterior, and encloses a small portion of fibrous matter. Each lobe is covered by pia mater, but not ensheathed by arachnoid as the nerves are; this latter membrane merely passes across from one side to the other along the inferior surface of the lobes.

The bulb of each olfactory lobe contains a small cavity, which may be demonstrated on a perfectly fresh brain by making a longitudinal section, with a very sharp knife, along the middle of the bulb. By directing a gentle stream of water upon either cut surface, a horizontal fissure will be laid open, extending nearly the entire length of the bulb, and placed nearer its inferior than its superior surface. This is the *ventricle* of the olfactory bulb, which is largely developed in the lower animals, the sheep, horse, ox, &c. and, according to Tiedemann, in the human fœtus at an early period.

By its posterior extremity the olfactory process is continuous with the brain in front of the locus perforatus. Its connection is established by one grey, and two white roots. The first may be distinctly traced from the posterior extremity of the olfactory fissure to be derived from the superficial vesicular matter of that part of the brain. This grey process may be followed along the upper margin of the lobe until it expands in the olfactory bulb. Of the white roots, one is external, the other internal. The external is remarkable for its great length. In removing the pia mater from the margins and floor of the Sylvian fissure this root may be seen, as a fine white cord, passing outwards to the floor of that fissure, where it is brought into proximity with the extra-ventricular portion of the corpus striatum; but there is no satisfactory evidence that it forms any special connection with that body. Sometimes two or three fasciculi of white matter may be seen passing from the posterior extremity of the olfactory lobe into the fissure of Sylvius in the same direction as that just described. The internal white root is very short, and is connected with the grey matter immediately internal to the locus perforatus. Each of the white roots, before their junction, is encrusted by a sheath of grey matter.

The union of the internal and external fasciculi of fibres or roots with the process of grey matter *(grey root)* forms the small but prominent posterior extremity of the olfactory lobe, which is called the *pyramid*. To this succeeds a narrow contracted portion of the lobe, *its trunk*, which is prismatic in shape, triangular in section *(trigonum nervi olfactorii)*; the apex projecting into the fissure consists of vesicular matter; the base is fibrous. Along the inferior surface of the trunk of the lobe we notice a groove which commences at the pyramid and terminates at the bulb. This groove has been supposed to indicate the rudimentary state of the canal of communication between the olfactory ventricle and the lateral ventricle, which is fully developed in the lower mammalia, (sheep, horse, &c.) and which may perhaps exist for a very brief period in the human fœtus.

It is worthy of remark that the inferior surface of the olfactory lobes does not extend beneath that of the anterior lobe of the brain. The lodgement of each olfactory lobe in an appropriate fissure provides for this as well as against compression by the superincumbent mass of the brain. As the lobes pass forwards they converge, and their anterior extremities or bulbs are separated from each other only by the crista galli.

The proper olfactory nerves are small fasciculi, which are implanted in the olfactory bulb, and pass from its inferior surface more or less vertically through the foramina of the cribriform plate to be distributed upon the mucous membrane of the nose. Their number varies, according to Valentin, from thirteen to twenty-seven. They are not always symmetrical in point of number on each side of the crista galli. When this want of symmetry exists, Valentin states that the openings in the ethmoid bone appear in most cases smaller and more numerous on the left side than on the right.

The many points of difference, both in structure and connections, which may be observed between the olfactory processes and the nervous trunks connected with the brain, justify our regarding them, not as nerves, but as lobes of the brain from which the true olfactory nerves take their origin. These differences may be enumerated as follows:—

1. The olfactory processes contain grey matter, which is not the case with any true nerve. 2. As they pass from the brain they *converge;* the nerves *diverge.* 3. Each ends in a bulb, from which numerous small nerves spring; nerves emerge through their proper cranial foramina as trunks. 4. Each terminal bulb contains a ventricle: a feature quite unique if we regard the olfactory processes as nerves. 5. The structure of the olfactory processes is that of brain; they have none of the fasciculated disposition so obvious in the optic and other nerves. 6. The shape of the trunk of each process is quite different from that of any nerve, being prismatic, whilst all nerves are more or less rounded at their edges.

Comparative anatomy confirms the views above stated. In every instance the small olfactory nerves which escape through the foramina in the cranial floor are implanted in a lobe or process connected with and similar in structure to the brain.

In the sheep, horse, and ox this process has in its bulb a large ventricle, which communicates by a canal with the lateral ventricle of the same side.

II. *The optic nerves*, properly so called, spring from processes of the brain called *the optic tracts*, which meeting along the median line form a junction beneath the tuber cinereum which constitutes the *optic commissure* or *chiasma*. It is from this part that the nerves may be properly said to take their origin.

Each optic nerve is enclosed in a dense fibrous sheath, which is continuous with the sclerotic tunic of the eye. Within this sheath is the proper neurilemma of the nerve, which is of very delicate structure, and dips in between the numerous fascicles of fibres of which the nerve is composed. The fascicles are small but distinct, and pretty uniform in size, and a transverse slice of the nerve from which the nervous matter has been dissolved out, has, when dried, a cribriform structure owing to this arrangement of the neurilemma. The optic nerves, in their course from the chiasma, diverge from each other, and escape from the cranium through the optic foramina in the sphenoid bone.

It will be necessary to study in connection with this nerve, first, the course, relations, and structure of the optic tract, which may indeed be regarded as a portion of that part of the brain with which the sense of vision is associated; and, secondly, the structure of the chiasma.

Of the optic tract.—The examination of this portion of the brain may be commenced from the chiasma. Traced from this part, it appears as a flattened band of nervous matter with rounded margins, which increases in breadth as it passes backwards. It may, for facility of description, be divided into three stages. The first stage extends from the chiasma to the anterior edge of the crus cerebri; in this stage the tract approaches in shape nearly to that of the optic nerve; it is narrow, and its edges more rounded than in the other stages; its outer edge corresponds to the locus perforatus, and its inner margin to the tuber cinereum, which extends between the optic tracts. It has been supposed that fibres of the tract emanate from the tuber cinereum, but this is matter of considerable doubt. It is certain that several nerve fibres are mingled with the vesicular matter of this layer, but it seems more probable that they pass between the crura cerebri than that they enter the optic tract.

The second stage of the optic tract is that which passes beneath the crus cerebri; here the tract expands a little; its outer margin adheres firmly to the crus cerebri and is overlapped by the inner convolutions of the middle lobe, and its inner edge is free, having no adhesion whatever to the crus, so that the handle of a knife may be passed between it and the surface of the crus. At the posterior margin of the crus, the optic tract expands considerably and becomes flattened and attenuated; and here commences its third stage, where it passes into a recess overlapped by the posterior extremity of the optic thalamus, and bounded within and in front by the crus cerebri. At this part the fibres of the tract diverge and form connections with the vesicular matter of the gangliform bodies situated here. These are the corpora geniculata, the quadrageminal bodies, and the optic thalami themselves.

The anatomy of the geniculate bodies is of great interest in connection with that of the optic tract. The *internal* geniculate body, more prominent and generally larger than the external, is placed in the angle between the posterior free extremity of the optic thalamus and the crus cerebri. It has apparently a very slight connection with the fibres of the inner or concave margin of the tract, which seem merely to touch it. A very distinct band of nervous matter is continued from this geniculate body to the testes, which, however, has no obvious connection with the optic tract. The *external* geniculate body is placed external and rather anterior to the inner one, and seems more intimately connected with the thalamus than the latter. The outermost fibres of the optic tract very distinctly expand over it, and are evidently continued inwards in a curved course along the inferior surface of the projecting extremity of the thalamus to the nates. Although I am not prepared to deny a connection between the inner geniculate body and the inner fibres of the optic tract, there can be no question that the principal fibres of this tract are traceable to the external geniculate body and to the nates.

When these bodies are examined in the brains of some of the inferior animals, (the horse, ox, sheep, cat, dog, and monkey,) they will be found to retain the same relation to the optic tracts and to the optic thalami as in man; but from a difference in the position and developement of the thalami themselves, the geniculate bodies do not preserve the same relation *to each other* as in the human subject. Hence the anatomist may readily be led to regard that as the analogue of the external geniculate body which really corresponds with the internal, and some recent writers have fallen into this error. The *external* geniculate body is incorporated with the posterior extremity of the thalamus; its structural distinctness from the thalamus may, however, be readily demonstrated by a horizontal section. The optic tract expands very freely over it, embracing it, and then passes on to the anterior margin of the nates. The *internal* geniculate body (which is in reality external and posterior to the former) still occupies, as in man, the angle between the thalamus and crus, is of great size, touches the margin of the optic tract by its anterior edge, and is most distinctly connected with the testes by a rounded mass of nervous matter, which establishes a complete continuity between the two bodies.

That the optic tract forms an extensive connection with the

optic thalamus cannot be doubted. This takes place partly by the adhesion of its convex border to the thalamus, and partly by its intimate connection with the external geniculate body. In the inferior animals already named, the connection with the thalamus through the external geniculate body is still more remarkable than in man.

Of the optic commissure or chiasma.—In tracing the optic tracts in a direction opposite to that pursued in the preceding description, they will be found to take a curved course from behind forwards, crossing the crura cerebri, and converging in front of the tuber cinereum and mamillary tubercles. They meet and seem to be fused together along the middle line. A flat body, similar in external characters to the optic tracts, but somewhat wider than them, is thus formed. This is called the optic commissure. The term *chiasma* is also applied to it, from χιαζω, *decusso*, either from its external form, or from a very generally admitted peculiarity of structure, namely, the decussation of some of its fibres.

From the junction of the two trunks along the middle line, and the divergence of the optic nerves from the anterior margin of the chiasma, four angles result, an anterior and a posterior, a right and a left. The posterior border of the chiasma is concave, and is formed by the innermost fibres of the optic tracts, which are continued from one side of the brain to the other. These fibres bear no immediate relation to the optic nerves, and exist well developed when that nerve does not exist, as in the mole, in which animal they appear as a band of commissural fibres connecting the thalami of opposite sides.

The anterior border of the chiasma is composed of fibres which appear to have no connection with the tracts, but which form the inner margin of each optic nerve; they take a curved course, and are concave forwards. These fibres probably extend between the retinæ of opposite sides, and may be regarded as inter-retinal commissures. Between these two sets of fibres are the decussating fibres, which may be best demonstrated on a specimen which has been hardened in spirit. These cross and interdigitate with each other, one set coming from the right, the other from the left side, so that a considerable portion of the fibres of the left optic nerve comes from the right optic tract, and vice versa.

Such is the structure of the human chiasma, as seems most probable from the coarse dissection of it as hardened in spirit. Microscopic investigation throws no light on this structure, beyond developing the fact that it is composed of fibres of the same kind as those which exist in the optic tracts.

In many of the lower animals the decussation of the optic nerves is developed in a manner much more striking than in the human subject. In osseous fishes (the cod, halibut, &c.) the right and left optic nerves completely cross each other, the nerve to the right eye taking its origin completely from the left side, and that to the left eye arising from the right side. In most birds a great part if not the whole of the fibres of each optic tract decussate, by distinct lamellar fasciculi which interdigitate with each other, and a similar disposition prevails in reptiles. In the inferior mammalia the chiasma has an appearance very distinct from that which it presents in the human subject. It is much thicker and more prominent on its inferior surface than the corresponding portions of each optic tract, and suggests the idea that the chiasma contains within it additional nervous matter to that derived from the optic tracts. The disposition of the fibres within this tract, as represented by Müller in the horse, is, however, identically the same as in the human subject.

Besides the facts of comparative anatomy which tend to confirm generally the existence of true decussating fibres within the chiasma of the human subject, some interesting observations have been made upon cases in which atrophy of one optic nerve ensued upon disease or destruction of either eyeball. In the great majority of such cases the optic tract of the opposite side, not that of the same side, exhibited marks of wasting, denoting that the filaments of each optic nerve which are most employed in vision have their origin from the opposite side of the brain; but in some instances the optic tract on the same side with the atrophied nerve wastes, while that of the opposite side remains apparently unaltered. The explanation of these contradictory facts has yet to be ascertained.*

III. *The third pair of nerves* or *motores oculi* are seen at the base of the brain connected with each crus cerebri, and separated from each other by the pons Tarini. Each nerve may be traced into the corresponding crus. If the dissection be made on a recent brain and under water or spirit, the component fascicles of the nerve are seen to diverge as they penetrate the crus, and to break up into their component fibres in the locus niger, where probably the fibres form an organic connection with the processes of the large caudate vesicles which exist in great numbers in that vesicular matter.

IV. *The fourth pair of nerves*, called by Willis *nervi pathetici*, are the smallest of the encephalic nerves. From their minute size and delicate structure the dissector must use great caution in tracing them back to their origin. They are remarkable for their great length of course from their origin to the point at which they escape from the cranium. In this course the fourth nerve winds round from the posterior and superior surface of the mesocephale, above the crus cerebelli and outside the crus cerebri, where it is accompanied by the superior cerebellar artery: it then passes forwards to the posterior clinoid process of the sella Turcica,

---

* See figures illustrative of these instances in Dr. Mayne's excellent article "Optic Nerves," Cyclop. Anat. vol. iii. fig. 420.

and at that situation penetrates a canal in the dura mater which is external to that in which the third nerve is lodged.

The fourth nerve emerges from the superior surface of the mesocephale in close connection with the testes, either from them or immediately behind them. It comes, in short, from the olivary column, which extends upwards beneath the quadrigeminal tubercles. In some subjects the separation of its fibrils at its implantation in the mesocephale is very obvious.

V. *The fifth pair of nerves* are the largest of the encephalic nerves, and from their peculiarity of origin possess great physiological interest. The fasciculated character is exhibited in them to a very striking degree. Their neurilemma is coarse, and the component fascicles of the nerve are large and easily separable.

The fifth nerve emerges from the inferior surface of the crus cerebelli by two fasciculi, which are the two roots or divisions of the nerve. One of these is of considerable size, *portio major*, and is said to consist of from seventy to one hundred small fascicles; the other, *portio minor*, is very much smaller, and consists of not more than five or six fascicles.

Each of these portions emerges from the crus cerebelli at a fissure formed by a separation of the transverse fibres of the pons which form the inferior portion of the crus. The fissure from which the larger portion escapes is separated from that which transmits the smaller portion by means of a few fibres of the pons.

The smaller portion or root of the nerve is situate behind and above the larger root. As it proceeds onwards from its point of emergence it passes obliquely upwards to the superior margin of the larger root, and reaches the same foramen in the dura mater as the larger root, but passes through it beneath and concealed by that root.

The larger root of the fifth nerve experiences a marked constriction just at its point of emergence from the crus cerebelli, but quickly enlarges again. It passes forwards and outwards to a large foramen in the dura mater. This foramen is oval, with its long axis directed forwards and inwards. It is situate above the internal third of the petrous bone, and is crossed superiorly by the superior petrosal sinus.

Both roots of the fifth nerve pass through this foramen and enter a space situate between two layers of dura mater, placed above the superior surface of the petrous bone at its inner third, and crossed over by the superior petrosal sinus. Within this space the large root lies above the small and conceals it. Here too the large root swells into a large ganglion, of triangular form, with curvilinear base. This ganglion (*ganglion Gasserii*) is remarkable for its great size, the looseness and distinctness of the fascicles which enter it, and its plexiform character, its true nature being, however, unequivocally shown by the deposition of vesicular matter in its meshes. From this ganglion proceed the three divisions of the nerve.

Both the large and the small roots of the fifth nerve may be traced through the superficial fibres of the crus cerebelli into the central part of the medulla oblongata (the olivary column) down to its lowest part, where they connect themselves with special accumulations of vesicular matter. There is no evidence whatever to show that the smaller root has any connection with the fibres of the anterior pyramids, excepting such an intermediate one as through the connection of some fibres of the pyramids with the mass of vesicular matter in which it is implanted. If the roots of the fifth nerve be carefully traced in a recent pretty firm brain, or in one hardened in alcohol, each root will be found to exist as a distinct band of fibrous matter, until its implantation in the vesicular matter of the olivary column.

This disposition of the roots of the fifth nerve is very plain in the brains of some of the inferior animals, (the sheep, horse,

&c.,) where, in consequence of the less perfect developement of the crus cerebelli, the course of each root is less concealed from view, and requires less dissection to expose it in its entire course.

VI. *The sixth pair of nerves (abducentes oculi)* lie immediately behind the posterior margin of the pons Varolii. Each nerve appears to spring from the corresponding anterior pyramid, into the interior of which it may be traced. It is highly improbable, however, that the fibres of the nerve have any other relation to the pyramids than that of passing through them from some subjacent vesicular matter in the olivary columns, in the same manner as the third nerve passes between the fibres of the crus cerebri from the locus niger. Indeed the sixth may be, justly enough, viewed as a portion of the third separated from it by the intervention of the fibres of the pons.

VII. *The seventh pair of nerves* consist of two portions,—*portio dura—portio mollis,*—which, differing in structure, distribution, and function, have no other claim to be classed together than that of emerging from the medulla oblongata in close proximity to each other.

1. *Of the portio dura, or facial nerve.*—The two portions of the seventh nerve lie in a fossa, bounded on the inside by the upper extremity of the olivary body, behind by the restiform body, and above by the posterior border of the pons. The portio dura is internal and much smaller, and more cylindrical than the portio mollis, which is distinguished by its soft and delicate appearance and its flattened form.

The facial nerve sinks into the central substance of the medulla oblongata, into which it cannot be traced to any great depth, but where, no doubt, like the other nerves of this part of the encephalon, it connects itself with a special accumulation of vesicular matter. Its origin may be referred to this source, where it comes into close relation with the fifth nerve, so that impressions conveyed by this latter nerve to the centre may readily affect the vesicular matter in which the portio dura is implanted.

Arnold, Gædechens, and others have described the portio dura as taking its origin by two roots. They regard the nervous fibres which are intermediate to the portio dura and portio mollis as the smaller root. These fibres had already been described by Wrisberg as forming an anastomotic branch between the two nerves. They cannot be considered as forming an additional root to the facial, because they have no connection with the centre distinct from that which the principal fibres of that nerve have; they seem rather to form a *branch* of the facial which accompanies some of the ramifications of the portio mollis into the labyrinth, and is distributed to the extensive muscular apparatus existing in that part of the organ of hearing.

2. *Of the portio mollis, or auditory nerve.*—We can very readily distinguish two roots in this nerve. One penetrates to the central part of the medulla oblongata in the same way, and following the same direction as the portio dura, but passing to a much greater depth into its substance. The other winds round the corpus restiforme, not penetrating it, but simply adhering to its surface, until it reaches the floor of the fourth ventricle, where it connects itself with the olivary columns, and in many instances is evidently continuous with the white striæ on either side of the calamus scriptorius, which for that reason have been very generally regarded as fascicles of origin of the auditory nerve.

The auditory nerve lies immediately contiguous to the small lobule of the cerebellum, the *flock* of Reil, *lobule of the auditory nerve* of Foville. It accompanies the portio dura to the internal auditory foramen, at the bottom of which each nerve escapes from the cranial cavity.

VIII. *The eighth pair of nerves* consist of the three following, the glosso-pharyngeal, the par vagum, and the spinal accessory. All these agree in their mode of origin, which approximates very much to that of spinal nerves. They arise by several distinct fascicles of filaments from the side of the medulla oblongata immediately behind the posterior border of the olivary body, all emerging in nearly the same line or sequence, the intervals between them increasing as they approach the spinal cord.

1. *The glosso-pharyngeal nerve* is much the smallest of the three portions of the eighth pair. It consists of from three to five fascicles of filaments, which emerge, very close to each other, from the substance of the medulla oblongata, just behind the upper extremity of the olivary body. The component fascicles are collected into two, a large and a small one, which have been regarded by Müller as distinct roots.

More than fifty years ago Ehrenritter, and lately Müller, described a small ganglion formed on the smaller of these bundles and situate in the upper part of the foramen lacerum *(ganglion jugulare)*. Being limited to a certain number only of the filaments of origin of the nerve, this ganglion may be regarded as so far analogous to that on the posterior roots of spinal nerves.

2. *The par vagum*, or *pneumogastric nerve*, lies immediately below the glosso-pharyngeal, and is more than twice as large as that nerve. It consists of from eight to ten fascicles of fibres, which emerge from the central part of the medulla oblongata behind the posterior border of the olivary body. According to Stilling and Wallach the fibres of origin of this nerve are implanted in a distinct accumulation of vesicular matter. The fascicles converge and form a single trunk, which escapes with the spinal accessory nerve through an opening in the dura mater common to both.

3. *The spinal accessory nerve* is partly of spinal origin, as its name implies. Its roots are numerous and consist of small

fascicles, which emerge from the side of the medulla oblongata and of the upper part of the spinal cord. The roots which are uppermost take a horizontal direction; the lower ones are oblique.

A careful examination of the lateral surface of the spinal cord, from which these roots arise, shows that the lower roots of the spinal accessory nerve emerge from the substance of the cord, quite close to the posterior roots of the cervical nerves, and that all the roots are nearer the posterior than the anterior part, whether of the cord or of the medulla oblongata. At the same time it must be remembered that these roots emerge too far forward to warrant the supposition that they have any connection with the posterior horns of the grey matter in the cervical region.

The nerve which results from the union of these radicle fascicles ascends, inclining slightly outwards, through the occipital foramen, and escapes in common with the vagus nerve through an opening in the dura mater situate at the foramen lacerum posterius.

We must here notice that very frequently some roots of the spinal accessory nerve coalesce with the posterior roots of the sub-occipital and of the first cervical nerve; this seems to be no more than an accidental admixture of filaments owing to their propinquity of origin; the fact is therefore one of little or no physiological significance.

IX. *The ninth pair of nerves* in its mode of origin by several fascicles of fibres, resemble the subdivisions of the eighth, which we have just examined.

The fascicles of origin of this nerve are ten or twelve in number; they emerge from the medulla oblongata along the anterior border of the olivary body, and their true origin is no doubt from the olivary column. At their emergence they are quite distinct from the anterior pyramids.

These fascicles unite into two, which coalesce and emerge as one nerve through the anterior condyloid foramen. Sometimes this coalition does not take place till after each fascicle has passed through a separate opening in the dura mater, leading to the anterior condyloid foramen. Such an arrangement has an obvious resemblance to that of the roots of spinal nerves.

In concluding this brief account of the origins of the encephalic nerves, the following remarks, which are of interest in a physiological point of view, suggest themselves.

1. We observe from how small a portion of the centre these important nerves take their origin—nerves, which control so many vital functions. Nevertheless we find that this portion of the centre is rich in vesicular matter, which is disposed in separate masses for the special reception of separate nerves. (See *fig*. 37, p. 275.)

2. The relative position of certain of these nerves to each other has an obvious reference to their mutual physiological influence. Thus, the optic nerve is at its origin immediately related to the third and fourth pairs, and less directly to the sixth and to the portio dura of the seventh. The fifth nerve is in the immediate vicinity of the eighth pair. The portio dura of the seventh, besides being in close relationship to the fifth, is equally near to the portio mollis. Anatomy, therefore, plainly suggests that these nerves must exert an influence upon each other, which many familiar actions confirm. Thus, dashing cold water in the face excites the respiratory actions, and also throws into action the muscles of the face, through the influence of the fifth nerve upon the vagi and upon the facial nerves. A powerful light thrown suddenly into the eyes causes all those actions which denote dazzling. In these movements the muscles which are animated by the third, fourth, and sixth nerves, and by the portio dura, are concerned, and these nerves are excited through the influence of the stimulus of light upon the optic nerve.

3. The hemispheres of the brain have no *direct* connection with any of these nerves; nor in all probability have the hemispheres of the cerebellum. The olfactory nerves *appear* to constitute an exception to this statement; but such really is not the case. The true olfactory nerves are implanted in the grey bulbs of the olfactory lobes of the brain. These constitute the true centres of these nerves, and their connection with the hemispheres is through the medium of commissural fibres and of vesicular matter. The other nerves of pure sense present an analogous, although not a precisely similar arrangement. Thus the optic nerve is connected with the geniculate bodies and with the quadrageminal bodies, some or all of which may be regarded as its proper centre, in which its fibres are implanted, and beyond which they do not proceed further towards the brain. The auditory nerve finds its proper centre in the olivary columns. And there are good anatomical grounds for extending this view to all the encephalic and all the spinal nerves. Each has what may be termed its special *centre of implantation,* upon which the nerve fibres depend for the maintenance of their proper power. And it is through the connection of these centres of implantation with the centre of volition or of sensation, by fibres distinct from those of the nerve itself, that each of these centres themselves and the nerves implanted in it participate in mental nervous actions.

CHAP. XI.

Sketch of the microscopical anatomy of the spinal cord and brain.

We conclude our account of the anatomy of the spinal cord and brain with a rapid glance at the present state of our knowledge of their minute anatomy as revealed by microscopical observation.

The elements of the two kinds of nervous matter, *fibrous* and *vesicular*, have been already sufficiently described. We shall only remark here that the great object of the anatomist's research should be to find out the precise manner in which the nerve-fibres are united with the nerve-vesicles. Of their intimate connection there can be no doubt,—much less of the influence which they are capable of exerting mutually upon each other.

Among the peculiarities of the fibrous matter in the centres it may be here stated that the fibres pass through a much greater range of size than in the nerves; that here we meet with nerve-tubes of the largest size, and, on the other hand, with minute fibres which seem to be continuous with the branching processes of the caudate nerve-vesicles. These fibres are perfectly transparent and differ from the nerve-tubes in the absence of any of the white substance of Schwann, and of the tubular membrane.

Some idea of the relation of the vesicular and fibrous matter in different parts of the cerebro-spinal centre may be formed by examining thin sections of the several portions of them made in various directions. It is impossible to make these sections sufficiently thin to enable us to explore a large surface with a high power, for which great transparency is necessary. Such sections, however, may be examined with low powers, as Stilling and Wallach have done. It is important, however, to notice that the appearances observed in this way afford no *certain* indication of the course and direction of the nerve-fibres, nor of the situation of the finer elements of the vesicular matter. The nerve-tubes are too minute to admit of being followed with an object-glass which magnifies less than from two hundred to three hundred diameters; yet Stilling's researches have been made with a power of no more than ten or twelve diameters.

The fibrous matter of the spinal cord consists of some fibres which pass either in a vertical direction, or obliquely, taking a long course, and deviating but slightly from the parallel to the axis of the cord. The fibres of the posterior columns are the most obviously longitudinal, and those which lie quite on the surface of the antero-lateral columns follow very much the same direction. Among the elements of the grey matter, fibres are found in great numbers, the direction of which is probably oblique or transverse, as considerable portions of them may be seen taking such a direction when a piece of grey matter, cut transversely, is examined under the microscope.

The grey matter of the cord contains caudate and spherical vesicles imbedded in their usual granular matrix. They are found in the horns as well as in the commissure. The caudate vesicles are most numerous, and exist in the anterior horn and at the root of the posterior one. The remainder of the posterior horn and the gelatinous substance which is found at its posterior border, resemble very closely in structure the grey matter of the cerebral convolutions.

By examining thin transverse sections of the cord, carefully hardened by immersion in spirits, a good view of the relative disposition of the grey and fibrous substances may be obtained. Stilling has carried investigations of this kind to a great extent, and has published some beautiful plates, which are quite true to nature. *Fig.* 36 is copied from one of them.

Fig. 36.

Transverse section of human spinal cord, close to the third and fourth cervical nerves. Magnified ten diameters. (*From Stilling.*)

f, posterior column; *i i*, gelatinous substance of the posterior horn; *h*, posterior root; *l*, supposed anterior roots; *a*, anterior fissure; *c*, posterior fissure; *b*, grey commissure, in which a canal is contained, which, according to these writers, extends through the length of the cord; *g*, anterior horn of grey matter containing vesicles; *e*, antero-lateral column, from *h* to *a*.

It is impossible, however, to obtain any information from such examinations, except of the most general kind. On referring to the figure, the reader will perceive several lines, of the same colour and appearance as the central mass, radiating from each horn of the grey matter to the surface of the cord, and not only to its external surface, but to that of its fissures. At whatever part of the cord the section be made, whether on a level with the roots of the nerves or between their points of emergence, the same appearance of radiating lines is seen, and the radiation will be found to extend between the central grey matter and whatever part of the surface of the cord the pia mater comes into contact with.

Stilling and Wallach suppose that these lines are continuous with the roots of the nerves; that they are, in fact, nerve-tubes proceeding from the grey matter to form these roots. But this supposition seems quite untenable, for the following reasons: 1st, because these lines are met with in situations intermediate to the points of emergence of the nerves; 2dly, because they pass to situations, such as the surface of the fissures, from which no nerve-roots emanate; 3dly, because, if they were nerve-tubes, they could not be so distinctly seen with so low a power. It is much more probable that they may be processes of grey matter prolonged towards the surface, to which bloodvessels may pass from the pia mater, or simply bloodvessels passing from the pia mater to the grey matter. In some well-injected specimens, which Mr. Smee had the goodness to shew me lately, the bloodvessels were seen to take exactly the same direction and course as these lines.

Besides the nerve-tubes which are found in considerable numbers in the grey matter, the branching processes of the caudate vesicles are met with in it also, which may be distinguished from the nerve-tubes by the absence of the white substance of Schwann, by their greyish colour, by their branching, and by their minutely granular texture. Capillary bloodvessels are met with in great numbers, ramifying in the grey matter, where they are much more numerous than in the fibrous matter.

Stilling and Wallach describe a canal passing through the centre of the grey commissure, and extending the whole length of the cord. This is certainly visible in most regions, but not in all. It seems to me to have much more the appearance of a bloodvessel than of a canal. According to these authors, it is the persistent condition of the much-talked of canal of the spinal cord referred to at a previous page. Its situation in the grey matter seems rather opposed to this view. The point, however, is one upon which I am not prepared to express a decided opinion at present, and which deserves more extended careful examination.

From a review of the preceding statements, it is plain that a large number of fibres pass into the grey matter of the cord, and probably form some intimate connection with its minute elements; and this fact is favourable to the supposition that the spinal nerves derive their origin, at least partly, from the grey matter. It must be admitted, either that these fibres unite in some way with the vesicles of the spinal grey matter, or that they pass through it up to the brain, in the grey matter of some part of which they become implanted; the former seems the more reasonable supposition, and more consistent with the apparent oblique or transverse direction which the fibres take in the grey matter.

The minute structure of the medulla oblongata resembles in many particulars that of the spinal cord. There is not, however, so complete an isolation of the fibrous matter in it as in the latter. Excepting in the anterior pyramids, and quite on the posterior and lateral surfaces, the two kinds of nervous substance freely intermingle. The anterior and posterior pyramids and the restiform bodies consist, at least in great part, of longitudinal fibres, but the remainder of the fibrous matter appears to be made up of transverse or oblique fibres. Most of these are doubtless connected with the roots of the many nerves which arise from the medulla oblongata. Stilling refers to special accumulations of vesicular matter connected with the roots of each nerve, and which probably form its proper origin. These contain large vesicles. It is impossible to give an exact interpretation to all the parts which are seen by his method of examination, imperfectly defined as they are from the use of such low magnifying powers. It would be waste of time and space to do more than refer to the representation given by Stilling (*fig.* 37) of the structure, as viewed by a magnifying power of ten diameters. Nothing can be more true to nature, so far as it goes, but its correct explanation must be sought for by diligent investigation with high powers. Numerous bloodvessels penetrate the central matter of the medulla, and no doubt many of the lines, which Stilling supposes to represent fibres, are in reality vessels passing to the grey matter.

The mesocephale has very much the same kind of structure as the medulla oblongata; transverse fibres (those of the pons) at its anterior part, longitudinal ones just behind these (pyramids), with vesicular matter freely intermixed. Its posterior part is the same in structure as the optic thalamus, and consists of numerous fibres with an abundant quantity of grey matter. The inferior layer of the crus cerebri is purely fibrous; its superior portion is identical in structure with the optic thalamus, and the locus niger contains large caudate nerve-vesicles, with a considerable quantity of pigment contained in them.

Microscopic investigation has as yet thrown no light on the *direction* and *connections* of the fibres of the cerebrum or cerebellum. What is known upon these points is derived from coarse dissection. The tubular fibres of which the white matter

Fig. 37.

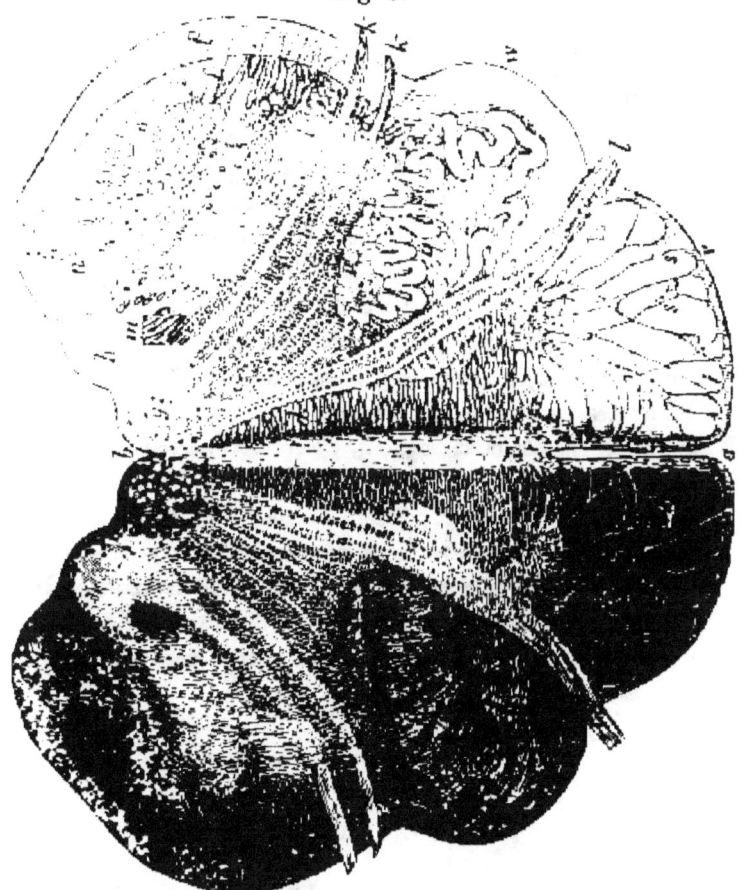

Transverse section of the medulla oblongata through the lower third of the olivary bodies. (From Stilling.) Magnified ten diameters.

, anterior fissure; *b*, fissure of the calamus scriptorius; *c*, raphé; *d*, anterior columns; *e*, lateral columns; *f*, posterior columns; *g*, nucleus of the hypoglossal nerve, containing large vesicles; *h*, nucleus of the vagus nerve; *i, i*, gelatinous substance; *k, k*, roots of the vagus nerve; *l*, roots of the hypoglossal, or ninth nerve; *m*, a thick bundle of white longitudinal fibres connected with the root of the vagus; *n*, soft column (*Zartstrang*, Stilling); *o*, wedge-like column (*Keelstrang*, Stilling); *p*, transverse and arciform fibres; *q*, nucleus of the olivary bodies; *r*, the large nucleus of the pyramid; *s, s, s*, the small nuclei of the pyramid; *u*, a mass of grey substance near the nucleus of the olives (*Oliven-Nebenkern*); *u, q, r*, are traversed by numerous fibres passing in a transverse semicircular direction; *v, w*, arciform fibres; *x*, grey matter near the root of the vagus.

is composed, appear to be disposed on different planes, and perhaps interlace with each other, so as to render it difficult to isolate any one of these planes to a great extent. This arrangement is more obvious in the cerebellum than in the cerebrum. The grey matter of both these segments contains the ordinary elements, caudate and spherical vesicles; but in the cerebellum those of the latter variety are much larger and more distinct than those which are met with in the brain. The peculiar structure of particular parts, as the optic thalami, corpora striata, tuber cinereum, &c. has been already described.*

The grey matter of the convolutions of the brain presents the same characters throughout, excepting in certain convolutions of the posterior lobe near the posterior and inferior horns of the lateral ventricles. Here, we may observe, in a horizontal section, the grey matter of the convolutions separated into two portions by a delicate white line, well represented in *fig.* 38. This layer of light matter was first described by Vicq d'Azyr, but has attracted very little attention from subsequent

Fig. 38.

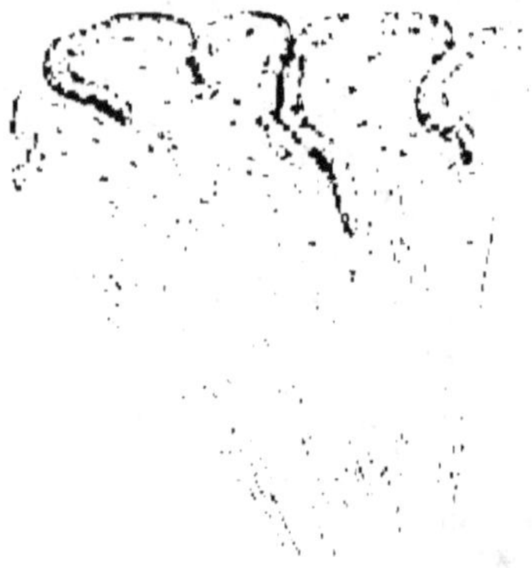

White line in the grey matter of convolutions of the posterior lobe of the brain.

* Vide pp. 225 & seqq.

anatomists. I have never looked for it without finding it. It consists of nucleated particles, similar to those in the grey matter of the cerebellum. The layer of grey matter external to it contains few nerve-fibres; that internal to it contains them in great numbers, passing into it at right angles. Any more complicated disposition of the grey matter of the convolutions, such as has been recently described by Baillarger and others, seems to me to depend not on a variety in the nature of the vesicular matter, but on some peculiar disposition of the vascular plexus causing a dark colour on one plane and a light one on another.

CHAP. XII.

An hypothesis of the action of the brain.

It is not intended to discuss here the physiology of the brain. But in order to develope more clearly than can be done in a mere description the connection of its several parts and the views of its structure which I believe to have the best foundation, I shall state briefly what appears to be the probable *modus operandi* of the organ, whether as the source of voluntary action or as the recipient of sensitive impressions.

It will be necessary first to state the following propositions as *postulates*.

1. The vesicular matter is the source of nervous power. In mental actions it is the part immediately associated with changes of the mind: whether in the working of the intellect, or in the exercise of the will, or in the perception of sensitive impressions.

2. The convolutions are the parts immediately concerned in the intellectual operation.

3. The simple exercise of the will, for a voluntary movement, is probably connected with the corpora striata.

4. The mere reception of sensitive impressions is connected with the optic thalami and the superior layer of the crus cerebri.

5. Mental emotions affect the posterior and superior part of the mesocephale.

6. The cerebellum is the regulator of the locomotive actions.

These propositions, which, it is admitted, although not improbable, are far from being proved, will serve as the basis of an hypothesis of the action of the brain.

In simple operations of thought, as in the exercise of the reasoning powers, or of those of the imagination, the convolutions of the brain are immediately engaged. We do not say that material changes give rise to the mental actions, but rather that the changes of the immaterial mind and those of the vesicular matter of the convolutions are simultaneous.

If an intellectual act excite the exercise of the will, the change in the superficial vesicular matter is propagated by the fibres of the hemisphere to the corpus striatum, whereby the will is excited, and the change in the vesicular matter of that body is propagated along the inferior layer of the crus cerebri, and, after passing through the mesocephale, along the anterior pyramids to the spinal cord, each anterior pyramid acting upon that antero-lateral column of the cord which is on the opposite side of the body to itself.

The pyramids connect the vesicular matter of the corpora striata with that of the spinal cord; from their small size it is highly improbable that they can be viewed as continuations of spinal nerves up into the brain.

A simple solution of continuity of the fibres of the hemispheres, which does not cause pressure, nor affect in any way the corpora striata, would therefore merely cut off the communication between the seat of intellectual action and the centre of voluntary action. The will, although unaffected, is unable to keep up with the train of thought, and mental confusion is the result. The loss of speech, which sometimes precedes a paralytic attack, and which may remain even after the paralysis has been removed, may be accounted for in this way. The intellect is competent to shape the thought, but unable to excite the will, upon which the exercise of the organs of speech is so obviously dependent.

Changes, originating or excited in either hemisphere, may be propagated to the corresponding parts of the other hemisphere by the transverse commissures, the corpus callosum, anterior commissure, &c. How far both hemispheres are in simultaneous action, during the rapid changes of the mind in thought, can scarcely be determined; it seems probable, however, that, in certain acts of volition, one only is the seat of the change which prompts to the movement. If I will to move my right arm, the change by which that movement is prompted belongs to the left hemisphere and corpus striatum.

Certain cases of disease confined to one hemisphere, in which a considerable degree, at least, of intellectual power persists, denote that the sound one may suffice for the manifestation of the changes connected with thought, and it may be reasonably supposed that the sound hemisphere may excite to action the centre of volition (corpus striatum) on the diseased side.

The existence of hemiplegic paralysis, then, implies an affection, *direct* or *indirect*, of the centre of volition (corpus striatum) on the opposite side. Pressure, or a morbid change in the physical state of its tissue originating in it or propagated to it, is all that is necessary for this purpose; and it is important to bear in mind that this change, like the change which takes place during the normal actions, may be of such a kind as to elude our means of observation.

When a sensation is excited, the stimulus acts from periphery to centre. The change is propagated by the sentient nerve to the optic thalamus, which, by its numberless radiations and its many commissures, is well calculated to excite all parts of either hemisphere, and even of both hemispheres. When the nerve excited is one of pure sense, the change is wrought more directly in the brain; if it be the fifth, or any of the nerves of the medulla oblongata, the stimulus acts directly on that part; but if a nerve of either limb be stimulated, the change must be propagated through the spinal cord.

It will be asked, if this be the *modus operandi* in sensations, how does it happen that disease of one optic thalamus does not impair sensation in one-half of the body? And how is it that such disease is much more frequently accompanied by hemiplegic paralysis, of a kind not to be distinguished from that which depends on diseased corpus striatum? The answer to the first question is as follows. The optic thalamus, or more properly, the centre of sensation, is never wholly diseased, for this centre is not confined to the optic thalamus of descriptive anatomists, but extends to the mesocephale and olivary columns. Extensive disease of this centre would probably be fatal to sensation. But the most ample provision exists for opening up new channels of sensation if those on one side or a part of them be impeded. The centres of opposite sides are intimately connected, especially in the medulla oblongata and mesocephale, by commissural or by decussating fibres; the optic thalami of opposite sides are connected to each other by the posterior commissure and the soft commissure, and the immense multitude of fibres which radiate from each thalamus insure its connection with a considerable extent of the brain, so that a change in any part of it cannot fail to be communicated to some portion of the hemisphere. It is sufficient for mere sensation that the centre of sensibility should be affected. Intellectual change resulting from that affection depends upon the fibres which radiate between that centre and the convoluted surface of the hemispheres (the centre of intellectual action).

It often happens that at the onset of a cerebral lesion *sensation* as well as motion is paralysed in the opposite side of the body.

In a few days, however, the sensibility returns whilst the paralysis of motion remains,—a fact which is sufficient to show that the motor and sensitive power must have different channels in the centres as well as in the nerves. The primary paralysis of sensation may be due to a lesion on one side affecting the centre of sensibility, or to the shock which that centre may have received from the sudden occurrence of lesion in some other neighbouring part. In the latter instance the recovery of sensibility takes place evidently on the subsidence of the effects of shock : in the former it may depend on the existence of other channels of sensitive impressions, independently of those involved in the lesion. Hence there may be lesion of one optic thalamus without loss of sensibility.

The answer to the second question is obtained from considering the intimate connection of the corpus striatum and optic thalamus. No two parts of the brain are so closely united by fibres passing in vast numbers from one to the other. Disease of the thalamus therefore may excite a morbid state of the corpus striatum, without producing any change in its structure, which may be recognised by the ordinary means of observation. And thus hemiplegia will take place, and remain as long as the morbid state of the corpus striatum remains. A lesion of the corpus striatum may in a similar manner affect the optic thalamus of the same side; but as that is not the only channel of sensitive impressions, a loss of sensibility does not necessarily occur from such a lesion.

Emotions are for the most part excited through the senses, A tale of woe, a disgusting or painful spectacle, a feat of wonderful power or skill, the sudden appearance of a person not expected, are calculated to produce corresponding emotions of pity, disgust or pain, wonder or surprise. But emotions may likewise be produced by intellectual change. The workings of the conscience may remind one of some duty neglected or some fault committed, and the emotion of pain, or pity, or remorse may ensue. Now emotion may give rise to movements independently of the will. The extraordinary influence of emotion on the countenance is well known, and this may affect one side of the face, which is paralysed to the influence of the will, or it may excite movements of the limbs, even when the will can exert no controul over them. From these facts it is plain that that part of the brain which is influenced by emotion must be so connected that the convolutions may affect it or be affected by it; that it may be readily acted on by the nerves of pure sense; that it may influence the spinal cord and the motor nerves of the face when the ordinary channels of voluntary action have been stopped. No part possesses these conditions so completely as the superior and posterior part of the mesocephale, which we have already noticed as concerned in acts of sensation. Is an emotion excited by an impression made upon one of the senses? this part becomes directly affected, and through the optic thalamus the emotional feeling causes intellectual change. The working of the intellect on the other hand may act on the seat of emotion through the same channel. And an excitement of this part may produce movement of a limb, or of all the limbs, by its influence on the spinal cord through the olivary columns.

The cerebellum influences the antero-lateral columns of the cord, partly through the deep fibres of its great commissure, the pons Varolii, which interlace freely with the fibres of the anterior pyramids, vesicular matter being interposed, and partly through those portions of the restiform bodies which penetrate the antero-lateral columns of the spinal cord. It associates and harmonizes the movements of the trunk, and especially those of the lower extremities, for locomotion, through those portions of the restiform bodies which are continued with the posterior columns of the cord.

The crossed influence of deep lesion of either hemisphere of the cerebellum is difficult to explain in the absence of any proved decussation of the restiform bodies. The connection of the deep fibres of the pons, however, with the anterior pyramids in the mesocephale does afford some explanation. If, for instance, the left cerebellar hemisphere be the seat of lesion, these fibres will be affected, and they may influence the fibres of the left pyramid, which again will affect the right half of the cord and the right side of the body. Those fibres of the restiform bodies which incorporate themselves with the antero-lateral columns, are doubtless too few to produce much influence.

FINIS.

www.ingramcontent.com/pod-product-compliance
Lightning Source LLC
LaVergne TN
LVHW012121120725
816032LV00019B/271